Praise for *Breaking the Gas Ceiling: Women in the Offshore Oil & Gas Industry*

"From fighting to control burning fields in a war devastated Kuwait to establishing industrial safety practices worldwide, who would have thought that women were key to these fundamental achievements? Ponton makes a significant contribution, as she gives voice to these outstanding professionals and gives credit where it is due. Too often still, the contributions and achievements of women professionals in the oil and gas sector go unaccounted. Breaking the Gas Ceiling gets the record straight, with elegance and substance.

Lourdes Melgar, Ph.D,
Robert E. Wilhelm Fellow, Center of International Studies (MIT)
Former Deputy Secretary of Energy of Mexico for Hydrocarbons and
Electricity

As a CEO, I believe it is imperative for today's generation of young women to realize there is a seat for them in the boards of oil & gas companies as the "gas ceiling" can be broken quicker and easier than before. Reading this book, they will think about these women who have gone before them and broken down those barriers in order to give them new opportunities.

Maria Moræus Hanssen, CEO, DEA Deutsche Erdoel AG

Everyone needs role models – and role models that look like you are even better. For women, the oil and gas industry has historically been pretty thin on role models for young women to look up to. Rebecca Ponton has provided an outstanding compilation of role models for all women who aspire to success in one of the most important industries of modern times.

Dave Payne, Chevron VP Drilling & Completions

Stories of successful women provide essential role models for an industry with so few in executive leadership. My advice is to be inspired and aim high.

**Melody Meyer, President Melody Meyer Energy LLC,
Women With Energy LLC, and NED**

My belief is that diversity is key to both creativity and solid long-term business results. Even in a country like Norway, where professional gender diversity is greater than in any other country I have had interactions with, we have an underrepresentation of women in top management positions. I would therefore like to express my appreciation to Rebecca Ponton for keeping this important subject on the agenda by presenting to us positive, impressive and, at the same time, obtainable role models.

Grethe K. Moen, CEO & President, Petoro AS

Rebecca Ponton has captured the compelling stories of many women, both the early pathfinders in the oil and gas industry and new entrants. Through these stories, it is very satisfying to now see that the industry has matured to be a place where anyone – man or woman – who commits themselves to high performance can succeed. No doubt we are all the beneficiaries of these intrepid women who have defined themselves by their work ethic and commitment.

Greta Lydecker, Managing Director, Chevron Upstream Europe

I applaud Rebecca Ponton's interest in increasing involvement and recognition of women as petroleum professionals.

Bruce Wells, Founder, American Oil & Gas Historical Society

As the industry now is more complex and faces more uncertainty, women will be more important contributors, especially in management and communication. Women could be just what is needed!

Karen Sund, Founder, Sund Energy AS

Women are essential in a high-technology environment such as the oil and gas sector. The complexity of this industry needs the diversity and parity of gender in order to be well-comprehended and managed. As one of the pioneers, I would very much encourage everyone who works in or wants to join this industry to read this book!

**Anne-Christine Dreue, Vice President Business Development EMEA/
Fokker Aerostructures B.V.**

The women featured in Rebecca Ponton's book have started to leave a trail where there was no real path and have found their place in the men's world of oil and gas. I encourage young women who want to join them to read this excellent book and follow the steps on the way to cracking the glass ceiling of the highly technical and complex oil and gas industry.

**Mireille Toulekima, Founder and Managing Partner,
MT Energy Resources**

This book is going to help create synergy among women in energy. Thanks, Rebecca Ponton, for being an agent of change.

Kimberly Smith, Radio Host, *Ask the Permian Landgirl*, FM107.1

In an industry with so few female role models, it's exciting to finally get to read a book about inspiring women who have carved out extraordinary careers in the oil and gas industry. None of these women's careers "just happened" – they were all borne of an untiring will to succeed and break the mold of traditional female roles. Every story tells of the incredible effort required to persevere through adversity and the personal and lifestyle sacrifices this has usually placed on them. This is a must-read for young women who have a passion to succeed in a male-dominated industry.

Amanda Barlow, Author
***An Inconvenient Life: My Unconventional Career as a Wellsite
Geologist***

This book is certainly well timed and will tell a story that needs to be out there.

Rebecca Winkel, Economic Analyst, American Petroleum Institute

Each woman's fight illustrated in this inspiring book is a fight on behalf of others to continue breaking barriers and ceilings. Rebecca Ponton has crafted role models for young women to be inspired by and to learn from. Each story is a story of passion, hard work, excellence and success, which makes this book a necessity, and a roadmap for women aspiring for a career in oil and gas.

**Laury Haytayan, MENA Senior Officer,
Natural Resource Governance Institute (NRGI) - based in Beirut**

Inspirational! A must-read for anyone, woman or man, in the industry. From chapter to chapter, you will be amazed at the courage and grit displayed in each story.

Ally Cedeno, Founder WOMEN OFFSHORE, LLC

This is an important book, relevant for our times. It shows that despite all sorts of challenges these women have succeeded in their professions of choice – in the offshore oil and gas industry. All of them have shown that through their belief in themselves, their competence and being strong enough to face a sometimes unwelcoming culture they have added valuable diversity of thinking, experience and leadership to our worldwide industry. I thank them all and hope this book will lead to many other women choosing to bring their own and unique character to this rewarding and important workplace.

Ms. Erica Smyth AC FAICD FTSE
Chair, National Offshore Petroleum Safety and Environment
Management Authority (NOPSEMA) – Australia
Co-author *Red Dust in Her Veins: Women of the Pilbara*

Rebecca Ponton's book serves as a great mentoring tool to inform women within the energy sector not only of opportunities to grow their careers, but also how to seek out and create those opportunities for themselves.

Joan Eischen, Director Markets at KPMG, Author
Energy and the City: Career Advice from Houston's Energy Executives

Rebecca Ponton vividly captures the triumphs and ordeals of a globally diverse group of highly accomplished senior women who have worked offshore. She combines those cameos with interviews with an equally varied set of entry and mid-career women, who have demonstrated great determination and have positioned themselves to be future stars. Together these vignettes provide valuable insights on how to achieve your own version of success.

Eve Sprunt, 2006 President of the Society of Petroleum Engineers,
Author *A Guide for Dual-Career Couples*

Rebecca Ponton has reached a new level of sharing about what is the inner core of women's leadership, by masterfully compiling stories of (very!) successful women in oil and gas, which are told in their own voice. The center of their achievements or challenges comes to light in a rich storytelling that is inspiring and worth reading!

Maria A. Capello, Awarded Energy Advisor
Lead Author of *Learned in the Trenches*

While I have waded through the 100 years of data and stories of female geologists in petroleum, I became aware of the paucity of women particularly in engineering. This book excites me about the future of women in this role and it is a delightful read.

Robbie Rice Gries, President, Priority Oil & Gas LLC
President of the Geological Society of America,
Author *Anomalies: Pioneering Women in Petroleum Geology*

The time for this book is now! We need to memorialize women's stories and make sure they're told.

Katie Mehnert, Founder/CEO Pink Petro,
Author *Growing With the Flow*

Breaking the Gas Ceiling:

Women in the Offshore ♀il & Gas Industry

Rebecca Ponton

Modern History Press

Cover photo Amelia Behrens Furniss (circa 1920), courtesy Noel Furniss.

Library of Congress Cataloging-in-Publication Data

Names: Ponton, Rebecca, 1961- author.
Title: Breaking the gas ceiling : women in the offshore oil & gas industry /
  by Rebecca Ponton ; foreword by Marie-José Nadeau, C.M.
Description: 1st Edition. | Ann Arbor, MI : Modern History Press, [2019] |
  Includes bibliographical references and index.
Identifiers: LCCN 2019012364 (print) | LCCN 2019015566 (ebook) | ISBN
  9781615994458 (Kindle, ePub, pdf) | ISBN 9781615994434 (pbk. : alk.
paper)
  | ISBN 9781615994441 (hardcover : alk. paper)
Subjects: LCSH: Women executives--United States--Biography. | Offshore
oil
  industry--United States. | Offshore gas industry--United States.
Classification: LCC HD6054.4.U6 (ebook) | LCC HD6054.4.U6 P66 2019
(print) |
  DDC 622/.33819092520973--dc23
LC record available at https://lccn.loc.gov/2019012364

NOTE: All names, ages, positions, titles, and roles were current at the time in which the interviews were conducted. Also, every effort has been made to give attribution to supporting documentation, to strive for accuracy, and to fact check.

Published by
Moder History Press
5145 Pontiac Trail
Ann Arbor, MI 48105

Distributed by
Ingram (USA/CAN/AU)
Bertram's (UK/EU)

www.ModernHistoryPress.com
info@ModernHistoryPress.com

Tollfree (USA/CAN/PR) 888-761-6268
FAX 734-663-6861

In loving memory of my mother,

Lois Francisco Lester.

You were a pioneer in your own right and always my biggest champion. I aspire to be more like you.

And my youngest brother,

Kirk Vandervoort.

Your time with us was too short but you gave us the gift of your son, whom you loved "more than anything on God's green earth."

Contents

Foreword

"We could be concerned as much with sticky floors as we are with glass ceilings" Marie-José Nadeau (Alagos, 2015).

Recent history has been marked by women's changing role in both society and the workplace. At work, male fiefdoms have been cracked open with educated and ambitious women now playing an increasingly active part across a range of professions. Today, we see women doctors, lawyers, journalists and entrepreneurs rising up the ladder of their respective careers.

Sadly, the energy sector, especially the oil and gas industry, has lagged behind other industries. Working conditions in this industry are not easy; workers often have to face extreme heat or biting cold while carrying out physically strenuous operations in remote locations. Traditionally, this was considered to be a male domain but increasingly women are choosing to accept the challenge and are now playing an active part in key activities such as exploring and producing hydrocarbons for an energy-hungry world. As the remarkable women profiled in this book testify, the oil industry is being transformed step-by-step, woman-by-woman as individual women make personal career choices that take them to new frontiers.

This change is being driven by women themselves and the choices they make. The way women perceive themselves and their role at work has changed irrevocably as they opt for education pathways in science, technology and management. However, as they embark on their careers they still face resistance at every stage. Individual attitudes of men towards women have clearly improved but many corporations still lag behind in terms of policies and attitudes. This needs to change.

A report by McKinsey & Company (2015) states that women are still underrepresented at every level in the corporate pipeline. The report points out that:

> Many people assume this is because women are leaving companies at higher rates than men or due to difficulties balancing work and family. However, our analysis tells a more complex story: women face greater barriers to advancement and a steeper path to senior leadership.

There are hopeful signs of changing attitudes at the top of the corporate ladder. There is a widespread recognition that female leadership in organizations adds value and ensures optimum performance. Yet, as the McKinsey report points out

> ...based on the slow rate of progress over the last three years, it will take twenty-five years to reach gender parity at the senior-VP level and more than one hundred years in the C-suite. While CEO commitment to gender diversity is high, organizations need to make a significant and sustained investment to change company practices and culture so women can achieve their full potential.

The women profiled in this book are pioneers in their field. We need to break down barriers but above all we need to encourage more young women to follow their path. We cannot create leaders unless the pipeline is full. Not everyone has leadership potential and the pool needs to be larger if more women are to achieve a place in the C-suite. Fortunately, we are on the right track as this book so admirably shows.

Marie-José Nadeau, C.M.
Chair, World Energy Council (2013 – 2016)[1]

---

[1] Marie-José Nadeau was elected the first female chair in the 90-year history of the World Energy Council (WEC) in 2013. She was appointed a Member of the Order of Canada (C.M.) in December 2015. Source: World Energy Council

Preface

Shortly after becoming a petroleum landman[2] in 2011, I remember attending my first oil and gas conference. As I looked around the cavernous exhibit hall, I realized I was one of very few women. While I was not intimidated nor did I feel like I had stumbled into a place I didn't belong, it was still an odd realization. As I attended more conferences, exhibitions, and seminars, I continued to notice the same scenario. There were few women in attendance and even fewer female speakers and presenters. *Where are all the women*, I wondered?

Having been a journalist for 20 years before I became a petroleum landman, my curiosity was piqued. I started researching women's involvement in the oil and gas industry and was mystified by the lack of information. This was even more surprising to me because it was a female friend who had gotten me started as a landman and, in all my trips to various courthouses around central and south Texas, I had observed a fairly even split between male and female landmen. (Equally as interesting, as it was a career I started at midlife, was the fact that many women were of a "certain age," as we like to say euphemistically.) I quickly discovered that parity definitely was not the norm throughout the industry.

When I read a 2011 interview with the male CEO of an exploration and production company, in which he was quoted as saying, "I don't think I can name a CEO of an oil and gas company that is a woman," (Stonington, 2011), I was astounded! The more research I did, the more I came to think of these women as "invisible," even within the industry itself. Certainly they exist, but few people seem to know who they are.

This book is not about male-bashing (us vs. them) nor is it meant to be exclusionary – almost without exception, the women interviewed mentioned having a male mentor – but the numbers don't lie. Women

---

[2] In its simplest definition, "landman" refers to a man or woman who runs title to determine mineral ownership in a property.

make up nearly 47% of the workforce in America (and similar percentages in many other countries),[3] but according to a report by IHS Global, Inc. (2016) for the American Petroleum Institute (API), they comprise only 17% of the total employment in the combined oil and gas and petrochemical industries. The Petroleum Equipment & Services Association (PESA) Gender Diversity Study (Accenture, 2018) puts the figures at 15% of entry-level hires and 18% of experienced hires. Based on public awareness and exposure, you would think it is even less. It is important to hear more of these women's voices, see them quoted and interviewed in the media more frequently, and feature them at industry events, where often there is not a single woman among the guest speakers. Currently, there is a campaign to encourage diversity on speaking panels by eliminating "manels" (all-male panels).

Originally, when I began writing this book, I envisioned it as a broad overview of women's contributions to the industry. However, once I started interviewing women for the offshore chapter, it took on a life of its own and I realized it deserved to be a separate book, profiling some of the elite group of women who comprise a mere 3.6% of the offshore workforce (Oil & Gas UK, 2017). Pat Thomson, who worked as a materials and logistics supervisor until she was 72 – and dreams of returning offshore at nearly 74 – was the first woman I interviewed and it was her incredible story that was the inspiration for this book.

Not only is there a need to encourage more women to join the energy industry as a whole, there is just as great a need to create awareness of the women who have led and are continuing to lead the way. I am in awe of the women featured within these pages – many of whom have achieved firsts in their fields – and am so appreciative that they have entrusted me with their stories. I hope I have done them justice and that readers will be as inspired by them as I have been. While this book is not a comprehensive or exhaustive look at women in the offshore oil and gas industry, hopefully, it will be a step toward documenting their contributions to the genealogy of our industry.

It was difficult to decide how to arrange the interviews, as there is some chronological overlap in many of the stories. It also is not possible to present them according to region, as the oil and gas sector is a worldwide industry, and many women work outside their home countries in a variety of locations over the course of their careers, becoming world travelers in the process. And so we begin with Margaret McMillan, who,

---

[3] Source: US Bureau of Labor Statistics (2017)

at 95, is the matriarch of the book, and the creator of many of the water safety programs, which all offshore personnel are required to pass prior to going offshore, and which have greatly improved the personal safety of the intrepid women – and men – who work offshore. I want to close with Ocean Engineering major (Class of '17) Alyssa Michalke's interview because I believe she is representative of the face and voice of the new generation of the female offshore workforce.

Price Waterhouse Coopers (PwC; 2015) released a survey in which 57% of female millennials polled stated they would avoid working in a business sector that had a negative image, including the oil and gas industry, further underscoring the need to emphasize that women have been involved in this industry from the very beginning. *We are not new to the industry*. Women have been involved in the industry just as long as men have (although not in as great numbers). We are a minority but, until there is greater parity, we can be a vocal minority and we can be a visible minority.

It has been said, if we forget the past, we are destined to repeat our mistakes. I think it is equally important to remember the past so we can emulate our successes. I believe if girls – and women – are aware that they have had a history in the oil and gas industry, it will enable them to envision a future in it as well.

– *Rebecca Ponton*
February 2019

in·trep·id
inˈtrepəd/

adjective
adjective: **intrepid**

> fearless; adventurous (often used for rhetorical or humorous effect).

> "our intrepid reporter"

synonyms: fearless, unafraid, undaunted, unflinching, unshrinking, bold, daring, gallant, audacious, adventurous, heroic, dynamic, spirited, indomitable; Morebrave, courageous, valiant, valorous, stouthearted, stalwart, plucky, doughty;

informal gutsy, gutty, spunky, ballsy "our intrepid leader inspired us to forge ahead"

antonyms: timid

Origin: late 17th century: from French *intrépide* or Latin *intrepidus*, from *in-* 'not' + *trepidus* 'alarmed.'

Source: Google

<table>
<tr><td>1</td><td>**WOW – Women On Water:**
A Brief History of Women Offshore</td></tr>
</table>

The first woman ever to work offshore in the history of the petroleum industry is... a mystery.

"Offshore" being a relative term, Azerbaijan staked claim to the first offshore oil discovery in 1803 with the extraction of oil from two hand-dug wells 18 and 30 meters from shore in Bibi-Heybat Bay (Zonn, et al., 2010), so perhaps it could have been an Azerbaijani woman.

Although it would be almost 150 years later, an Azerbaijani woman did indeed make her mark on the industry. Maral Rahmanzadeh, already a respected artist, rose to greater prominence in the 1950s for her renderings of life offshore on Azerbaijan's famous *Neft Dashlari* (Oil Rocks) (Aliyev, 2011). Maral is said to have removed the traditional Muslim veil and donned a jumpsuit in order to work among the oilmen, capturing scenes of their daily lives (Nazarli, 2015). Perhaps not thought of as traditional "offshore work," art plays a vital role in documenting life offshore and is carried on by contemporary artists like Scotswoman Sue Jane Taylor profiled in this book.

While women have achieved many firsts in the field – and continue to do so – it is impossible to say with any degree of certainty who the *very* first woman was to work offshore. Because it is a worldwide industry, each country with a petroleum-centric economy would have had a woman who was the first to be involved in the industry whether she actually went offshore or remained onshore where she participated in some facet of the offshore industry.

Just as women were involved in the onshore industry from its inception in the mid-19th century when oil was first discovered in North

America in Oil Springs, Ontario, Canada, in 1858 and in the United States the following year with the Edmund Drake well in Titusville, Pennsylvania, we have to assume women were involved offshore, in some capacity, from the beginning.

American anthropologist Diane E. Austin, PhD, confirms this when she writes, "Long before the 1970s when oil and gas companies were forced by [US] federal civil rights laws and guidelines established by the Equal Employment Opportunity Commission (EEOC) to begin hiring women in offshore jobs, women were intimately involved with the industry" (2006, p. 173).

Despite that fact, not much was written about women working offshore in the US until the 1970s. Newspaper archives offer a fascinating glimpse into how the arrival (often referred to as the "invasion") of women affected the previously all-male bastion of offshore oil and gas.

A brief article appeared in some US newspapers in September 1973, stating five women became the "first of their sex" to work offshore in the Gulf of Mexico (GoM) when they were hired through a catering company to "take over cooking and cleaning jobs" on a drilling rig off the coast of Louisiana (no byline; 1973).[4] The article specified the women would work a week on/week off rotation schedule. There was no mention of the women's names nor were there any photos of the women to accompany the text – a strange phenomenon that still happens today.

A December 1975 *New York Times'* article profiled four women working on a rig in the GoM, but focused mainly on 20-year old roustabout Cindy Myer, nicknamed "Ralph" by her male co-workers, since she "work[s] like a man." Calling her "something of a pioneer," the article says female roustabouts offshore are "rare and controversial. And most of the controversy about them is back on land" (Sterba, 1975).

As Dorothy Mitchell the night cook on the rig pointed out, the opposition didn't always come from men, sometimes it was from their wives, as this March 7, 1977, letter in the iconic *Dear Abby* ("agony aunt") American advice column illustrates on the following page.

---

[4] While bylined articles first began appearing in US newspapers in the 1920s, they didn't become more commonly used until the '70s and often not even then. By the '80s, articles almost always carried a byline. (The author has noted the writer's byline whenever one is provided.)

Dear Abby:

Wife worries about hanky-panky with women on offshore oil rigs

DEAR ABBY: My husband works for an oil company, offshore seven days and onshore seven days. I'm just a housewife who can't even get to her own husband when he's working offshore, but listen to this, Abby. There are seven females who are now working side by side with the men on that rig, thanks to the government and Women's Lib!

Those women also eat and sleep under the same roof as the men. My husband says his company is bound by law to hire women, and there's nothing he can do about it.

There are plenty of jobs for decent women on land, so why would a decent woman want to work on an oil rig with a bunch of men? They say these women demanded equal rights. Where are MY rights?

My husband says I don't have to worry —that no funny business is going on and the men treat the women just like they were guys. Do you really buy that, Abby?—M. B. FROM TEXAS

DEAR M. B.: Yes, I buy it. And furthermore, any woman who works alongside a man on an oil rig is earning her bread the hard way. If she wanted to cash in on her femininity, I can think of several other jobs she could have chosen.

In 1977, former pro golfer Kristin Lovelace made news when she became the first woman operator on an offshore platform off the coast of California for Chevron. Surprisingly, in this instance, instead of an article, most newspapers ran one or two captioned photos of Kristin, saying she "works a regular shift along with men," (no byline, 1977), stating the obvious, but something that was an anomaly at the time.

Newspapers followed her career and two years later in June 1979, the *Los Angeles Times* included her in an article about Chevron's platforms off the coast of Santa Barbara, California, just miles from where the first offshore rigs in the US were erected in the 1890s (Japenga, 1979). This time, although the writer mentions she was "the only woman on the West Coast employed as a platform operator," the focus is more on Kristin's technical and mechanical abilities than the fact that she is a woman and, in the photo caption, she is referred to as a "trouble-shooting mechanical wizard." However, platform operator Don Finch reveals that, "A lot of guys got real upset," when Kristin first arrived offshore but says, except for a few diehards, they became convinced she was "obviously mechanically inclined" and capable of doing the same work they were.

It is common knowledge that in the early days (and the not-so-distant past) getting hired often was based on whom you knew. It may have helped that, according to the article, Kristin had been recruited by family friend, Chevron area supervisor Bill Ryherd, who at that point had been with the company nearly 30 years. Surely, some of his friends had sons he could have hired; apparently, he saw the potential in Kristin and was willing to go against convention.

"Women joining the ranks of offshore drilling" is the headline of an American newspaper article dated July 12, 1978, and sent over the wire by the Associated Press (AP) (no byline, 1978), which revealed, "A brief, unscientific survey of five platforms showed older men were concerned about the possibility of sexual encounters and the need to watch their language. Younger men spoke more about the competition."

Robert Moore, Professor of Sociology at the University of Aberdeen, in 1984 reported for *The Guardian* newspaper in London on an EOC-funded[5] study of recent female geology graduates seeking employment both onshore and off, and came to the blunt conclusion, "Exclusion from jobs by men is the problem" (1984).

Interestingly, things can work both ways and the workplace culture, whether offshore or on, is certainly influenced by the "top down" attitude. Even in the early days, occasionally there was a woman in charge. In an article from 1982, Sue Trolinger, then district manager of operations for Sun Exploration and Production's northern district in Tulsa, Oklahoma, whose job entailed working offshore, is quoted as saying, "I don't treat the men any differently on a rig than I do in the office" (Canetti, 1982).

Particularly in the early days, as Professor Moore's study found, and even more recently, the lack of accommodations has long been used as a reason why there aren't more women working offshore. Surely an industry capable of extracting hydrocarbons from beneath the ocean can figure out a way to accommodate all of the members of its workforce, regardless of sex. The writer of one article relayed a simple solution offered by a group of women working offshore in the GoM in 1978: "To the argument that a bed goes empty with a lone female on board, the ladies reply – put on another woman" (no byline, 1978).

---

[5] The Equal Opportunities Commission (EOC) was established under the Sex Discrimination Act 1975. In 2007, it merged with several other agencies to form the Equality and Human Rights Commission. Source: EOC

Sometimes, it only takes a little ingenuity – and a sense of humor – to solve the problem. Norwegian Anne Grete Ellingsen, the first female offshore platform manager (OIM) in the North Sea, recalls complaining about the open shower stalls when she was offshore in the early '80s. On her next rota,[6] as she walked toward the long row of showers, to her delight, she discovered one stall had been painted pink and covered with a pink shower curtain. "No man was going anywhere near that one!" she says, laughing at the memory. In fact, no man went anywhere near the shower area when she did. "I had it all to myself," she recounts gleefully.

Ann Cairns, now president of international markets for MasterCard Worldwide, and the first female engineer certified to work offshore on UK rigs in the '80s (Groden, 2015), is not alone in recalling having to sit with her feet against a bathroom door to ensure privacy. "There were no women's loos and no locks on the doors either" (no byline, 2013).

In a 2010 *Fast Company* article, chemical engineer Cynthia "C.J." Warner, now president & CEO of Renewable Energy Group, Inc., gave a vivid description of what passed for privacy offshore in the '80s, given the lack of women's accommodations, when she talked about being the only woman on a rig with 300 men and one bathroom – without a door. "You put the hard hat over your lap," she says with a grin (Kamenetz, 2010).

Women in countries like the US, the UK, Norway, and Australia now have a 40-plus year history in the offshore industry but, for women in other countries, the opportunities are only beginning to be within reach.

While the Middle East is synonymous with oil and gas, not all countries in the region have the same abundance of fossil fuels or the ability to exploit them. The offshore petroleum industry in areas like the eastern Mediterranean (Egypt, Israel, Lebanon, Cyprus), as well as others, is still in the frontier stages. As time goes on, the oil and gas industries of those countries certainly will see their own female pioneers.

Fresh discoveries are being made even in long-exploited areas such as the Gulf of Mexico and the North Sea. The French company Total, partnering with Chevron, announced its biggest find to date in January of 2018 in the GoM off the coast of Louisiana. Almost simultaneously, Royal Dutch Shell announced one of its largest finds in deepwater GoM about 200 miles from Houston, Texas. Within a 24-hour period, British Petroleum (BP) announced two new discoveries in the North Sea.

---

[6] Shorthand for "rotation schedule."

Despite what Gavin Bridge and Philippe Le Billon refer to in their book *Oil* as the "profoundly gendered nature of oil work" (2012), every new discovery – whether onshore or off – should be viewed as an opportunity for women to become more involved in the industry.

Although this book is not an in-depth exploration of the history of women offshore, instead focusing on individual women's accomplishments, particularly those who have achieved firsts in their fields, the importance of preserving this piece of history cannot be overstated.

As mechanical engineer Yassmin Abdel-Magied puts it so succinctly in her memoir, *Yassmin's Story,* "It can be disenfranchising to not be able to connect with your history" (2016). While she was talking about ancestry and heritage, her comment applies to history in a broader, more general sense and certainly to women in the petroleum industry. And, as the eminently quotable Yassmin wrote in a 2017 article for *Teen Vogue* online, "History matters, because it informs the attitudes of the present society" (2017).

To see the history of women in the industry portrayed in photos, scroll through the company history on the website of an oil and gas company and watch as the black and white photos from the early 1900s of men on rigs and in boardrooms progress to color photos around the 1960s and '70s of men on rigs and in boardrooms. More current photos that include women are often stock photos. To portray a more accurate picture (so to speak) of the history of women's involvement in the industry, company historians and archivists need to search for photos of female employees – they exist! No doubt there are women around the world who would be willing to share photos from their days offshore (and on).

While each of the women's stories in this book is unique, there are common threads that run through many of them – the love of a challenge and the willingness to rise to the occasion. Overcoming fears, whether they were specific to offshore, such as a fear of water or of flying or fear of the unknown – the uncharted territory they were entering. Determination in the face of obstacles. Resilience despite repeatedly being told no, encountering naysayers or being subjected to negative remarks. These are characteristics that seem to be innate to the women who have chosen to work offshore.

Paula Harris, global director corporate social responsibility programming at Schlumberger, went offshore in 1987 shortly after graduating from Texas A & M with a Bachelor of Science (BSc) petroleum engineering, and says in the seven years she worked offshore, "99.9% of the time, I was the not only the only female, but I was always the only person of color. Here I am, close to 30 years later, and I still run into people that remember me because I was always the odd man out. [It was] very different to find an African-American female offshore."

And when she did meet resistance, it helped to have a snappy comeback.

"Sometimes you get adversarial [comments], 'I don't want a woman out here!'

"And I used to love to say, 'Listen, you got the best. But if you want to send me back, it may be weeks before you get another engineer!'" (Career Girls, 2012).

So many of the women interviewed said they fell into their careers by accident or chance or serendipity, but they didn't just get lucky. It was – and is – their brave spirits and inquisitive minds that allowed them to succeed in uncharted territory and, in a twist on a pop culture reference, "To boldly go where no woman ha[d] gone before."

If there is anyone that embodies that spirit, it is an American woman named Amelia Behrens Furniss, who undoubtedly must have been one of the first women to work offshore and whose photo graces the cover of this book, thanks to the generosity of her grandson, Noel Mark Furniss, Retired Air Force Chief Master Sergeant, Air Force Central Command, who is the keeper of his paternal grandmother's photo albums and other memorabilia.

Amelia Florence Behrens was born July 6, 1895, just one year before the discovery of Summerland oilfield in Santa Barbara County, California, the site of the first offshore discovery in the US from wells drilled from piers that extended out to sea.[7] Her father, Captain Henry Behrens of Denmark, ran a commercial diving business before the turn of the 19th century and a diving bell exhibition on the Venice Pier in California, and taught both Amelia and her sister, Dorothy, to dive.

To say Amelia Florence Behrens Musser Furniss led a colorful life would be an understatement. In the Roaring Twenties and the era of the flappers, she was unpredictable, unconventional – and, yes, unflappable – seemingly up for any type of adventure. "The diving girl," as she was

---

[7] Source: County of Santa Barbara Planning & Development Energy Division: Summerland Oil & Gas Production

often called, first made it into the newspapers in March 1913 when her marriage at the age of 17 was listed in the *Los Angeles Times*.

In an October 1921 newspaper article, you can fairly hear the annoyance in the (presumably male) reporter's voice when he types, "If it isn't one thing, it's another with Florence Amelia Behrens Musser, woman diver. Florence Amelia manages to get on the front page of newspapers every few months" (no byline, 1921). He goes on to refer to her latest "exploit" – a dangerous descent into a 162-foot oil well through 24-inch casing to retrieve some tools – as if it were a media stunt. Amelia had been asked – or volunteered, depending on the account – to attempt to retrieve the expensive tools when Captain's Behrens' size prevented him from going into the pipe.

A faded tea-colored clipping in Amelia's scrapbook from the *Venice Vanguard Evening Standard* takes a different tone, with the front-page announcement of the triumph of the local "deep sea diver [who] has won a new record, unparalleled in the history of diving and underwater exploits" (no byline, 1921).

Months later, newspapers across the country were still running the story with the dramatic headline, "Sea Diver Fights For Life in Oil Well," but none ran Amelia's photo.

She made five dives over a period of three hours, in water 45 feet deep, setting an endurance record, but became wedged in the pipe on the final dive. By staying calm and keeping her wits about her, she finally was able to free herself. While the danger was real, a reporter for the *Oxnard Daily Courier,* writing in the sensational tone of the day, claims, "...she wondered whether she would ever come out alive" (no byline, 1921).

The story was given a small mention in the Comment section of *Petroleum Age*. Referring to her as "Miss Behrens" (although she was married) and the "daughter of Captain Henry Behrens, sea diver" the piece opens with the line, "Believe it or not..." and ends by saying, "We say this story takes the cake" (no byline, 1921). Was the incredulous tone because of the feat itself or because she was a woman or possibly both?

Despite extensive press coverage, it seems there were those who had their doubts that a woman had accomplished such a coup. The *Rig and Reel* magazine ran a photo of Amelia formally dressed in a coat and hat wearing dark lipstick next to the caption "*Somebody* said it couldn't be done – and so Miss Amelia Behrens did it" (no byline, 1922). It went on to say, "And others hinted that it wasn't done, so we asked Captain

Behrens to verify it – which he has done to our complete satisfaction." The magazine printed Captain Behrens' letter dated December 7, [year not given], in which he also wrote, "Amelia... is... perfectly fearless and has complete confidence in herself in whatever she does." And with that stamp of authenticity, it went on to show the photo of Amelia, which Captain Behrens verified was taken that day, her slight, 5'1", 118-pound frame encased in the 446-pound wet suit, along with an article detailing her record-breaking dive, retrieval of the tools, and timely escape.

As if hard hat diving weren't enough, Amelia also performed wing-walking and stunt work in silent films, such as the classic *Perils of Pauline* (1914). Noel recalls how Grandma Mimi would answer his "inquisitions" and when he asked her about the hard hat diving and wing walking, she told him she had to do those things because the men couldn't (Furniss, 2016).

Her intrepid, adventuresome spirit is embodied in the women who work offshore, and in 2012 the Women Divers Hall of Fame (WDHOF), an international non-profit professional honor society, recognized her achievements with her posthumous induction as a Pioneer Hard Hat Diver. Her legacy lives on through the Amelia Behrens Furniss scholarship set up by Noel through WDHOF to encourage young women to enter the diving profession. One young woman she has inspired is her great-granddaughter, Lindsey Noel Furniss, Noel's daughter, who is studying marine biology.

More recently, the 2016 movie *Deepwater Horizon,* based on the April 20, 2010, tragedy in the Gulf of Mexico, reminded the public that not only do women work offshore, they often fulfill critical roles. While the accuracy of some of the events depicted in the film has been debated, the movie gave the distinct impression that there was only one woman aboard the rig that night. That is often the case in real life, as many women in this book attest; however, the Transocean Personnel On Board (POB) list reveals there were actually five women on the Deepwater Horizon rig:

Oleander Benton, catering/baker

Andrea Anasette Fleytas, dynamic positioning officer II

Virginia Stevens, catering/laundry

Paula Walker, catering/laundry – days

Cathleenia M. Willis, mudlogger

There are a number of influential women whose stories aren't told in the limited amount of space available here, which would suggest that an encyclopedia of pioneering women in the offshore oil and gas industry needs to be compiled.

Dr. Marjorie Apthorpe

An Internet search of micropalaeontologist Marjorie Apthorpe's name will leave no doubt that she was one of the first Australian women to work offshore. Some articles assert that she was one of the first two (along with fellow Aussie reservoir geologist Judy Garstone) and specify that they went offshore in 1981. That story has been written and repeated so many times that it has become part of oil and gas lore. Marjorie herself had taken it as fact until she recently began going through some memorabilia of her mother's, as well as old papers and a photo album of her own from the 1960s, and recalled she had, in fact, first gone offshore in 1967 – fourteen years earlier than commonly reported!

"There were two trips to a Shell Development Australia rig (the SEDCO 135E) in the Otway Basin, Victoria. My feelings were of excitement, curiosity, a bit nervous riding in a helicopter for the first time." She also was disappointed not to be able to work offshore, but instead was given a tour of the rig, where she observed the work environment, and was shown equipment such as sidewall core guns and logging tools. On the second trip she wrote a sidewall core logging program. "I was only four years out of university and had not been exposed to much of the hardware and equipment side of exploration before, so it was definitely informative."

In December 1968, a year after Marjorie's first venture offshore, the *Melbourne Sun* newspaper ran a feature about her trip. She is still indignant when she recalls the headline: Woman in a Man's World: The Oil-Hunt Girl (Shand, 1968). The article, written by a female reporter, begins, "Meet Miss Marjorie Apthorpe, one of the few, if not the only woman in the world who has worked on an offshore drilling rig," and goes on to detail the complexities of Marjorie's work. While perhaps it wasn't necessary to conclude with "...she enjoys giving or going to informal parties and barbecues," it was a sign of the times. The rest of the article offers fascinating insight into the work of a micropalaeontologist (noting Marjorie was one of about 25 in all of Australia at the time) and is exactly the type of real-world success stories needed to illustrate the vast array of careers available to women in the industry.

Despite continuing her work in the oil and gas sector, it would be another 14 years before Marjorie went offshore again. She left Shell in 1973 to join Burmah Oil Company of Australia (BOCAL) as the micro-palaeontologist in Perth in the Woodside–Burmah joint venture. Burmah Oil sold its shareholding in 1976, and the operating company of the joint venture was rebranded a couple of times, and ultimately known as Woodside.[8]

According to Marjorie, "Both Burmah and subsequently Woodside had strict "no women offshore" policies. This was a source of extreme frustration for several women geologists on staff, who felt they could not achieve any sort of career path without that essential offshore experience." Marjorie and the other women approached the state government anti-discrimination commissioner only to discover that there wasn't any legislation that covered their situation.

The policy remained in effect until March 1981 when Marjorie says a crisis in an offshore well (in this case, not hitting the reservoir target at the predicted depth) required the company to send someone offshore who could tell from ditch cuttings where they were geologically and whether to continue drilling or abandon the well.

"That's why I, being the only company micropalaeontologist, was sent offshore. Judy Garstone, being a reservoir geologist, and the specialist geologist to work on cores if and when the well hit the reservoir, was sent with me on the basis that there was safety in numbers.

"Management were not happy at the time. I recall hearing about arguments between senior staff. By October 1981, things seemed to settle down because I was sent out to a rig for two weeks on my own. Presumably we had proved our worth," she says wryly. "However, during the second trip, I recall hearing that the rig boat captain was blaming *me* for the bad weather! Superstition dies hard" (Apthorpe, 2018).

Margareth Øvrum

In 1982, one year out of the Norwegian University of Science and Technology (NTNU), as a graduate civil engineer specializing in technical physics, Margareth Øvrum was hired by Statoil (now Equinor) and has remained with the company ever since, rising to her current position of executive vice president. At 32, she became the company's first female platform manager on the Gullfaks A, which at the time was

---

[8] Australia's largest independent oil and gas company. Source: Woodside

the world's largest offshore installlation. Not only did she often find herself the only female in a role, she was usually the youngest but, because of her skills and training – particularly her "time on the [rig] floor" – she had confidence in her abilities. Having broken that barrier, she saw "seven [or] eight female platform managers employed in a short space of time." As she has risen through the ranks of leadership, she has achieved her goal of seeing 40% of her senior management team comprised of women, but stresses that other factors, aside from gender alone, should be considered in the recruitment process (no byline, 2011).

Moira Ming Ying Li

A chemical engineer with a Bachelor of Engineering from the University of Strathclyde in Scotland, Moira Ming Ying Li joined Maersk Drilling in 2007. Maersk Drilling is a subsidiary of A.P. Møller – Maersk Group, the world's largest container shipping company.

In August 2014, she became Maersk's first female unit director and assumed her role on the Deliverer rig working for Chevron offshore Angola.

Maersk affiliates have seen a number of women hold C-suite positions: Gretchen Watkins was Maersk Oil's Chief Operating Officer (COO) in 2014 and would be named its Chief Executive Officer (CEO) in 2016 before going on to become Shell Oil's first female CEO in 2018. From 2015 to 2017, Ana Zambelli, who has a master's degree in petroleum engineering, held the position of Chief Commercial Officer (CCO) of Maersk Drilling. In yet another first, Carolina Dybeck Happe was appointed the first female Chief Financial Officer (CFO) in A.P. Moller – Maersk Group's 114-year history, effective January 2019.

Penny Chan Wei Chze

After graduating from the Universiti Teknologi Petronas (UTP) with a bachelor's degree in petroleum engineering, first class honors, and within four years of joining Petroliam Nasional Berhad (known as Petronas), Malaysia's government-owned oil company, 26-year old Penny Chan Wei Chze became its first woman deputy drilling supervisor. In a common scenario, she has found herself to be the only woman working among 150 men offshore and says, "To be effective on a male-dominated rig, I always put the gender aside, and only focus on professional strength" (Cheang, 2015).

Harshini Kanhekar

Having already earned a Bachelor of Science (BSc) and studying for an MBA, Harshini Kanhekar became the first woman to attend the National Fire Service College in Nagpur, India, in its 46-year history. Incredibly, a country that in 1966 named Indira Gandhi its first (and, to date, only) female prime minister had never had a woman firefighter until Harshini's graduation. After joining the Oil and Natural Gas Corp. (ONGC) in 2006, she eventually gained access to offshore rigs. "I am very thankful to my management who encouraged and believed that a woman could handle offshore drilling services. Until my posting, no woman was given the chance to serve [on] offshore rigs" (Aranha, 2017). She says, "Women today are doing the unthinkable." Making a point that it is societal mores and attitudes that need to catch up with women and not the other way around, she adds, "It doesn't imply that we couldn't do it ten years ago" (Nayak, 2016).

Griselda Cuevas

Griselda Cuevas of Mexico, who holds a master's degree in engineering and is fluent in four languages, worked offshore Angola for Schlumberger in 2008 and was, at that time, the only female engineer assigned to deepwater operations. After nearly two years offshore, she left the petroleum industry and became involved with tech start-ups, a prime example of how skills gained in the oil and gas sector are transferable to other industries, and has been employed with Google since 2014 (Cuevas, 2012).

Dena Hegab

Following in the footsteps of her father, a drilling engineer, Dena Hegab of Egypt earned her Bachelor of Science, Petroleum & Energy Engineering at the American University of Cairo. Within five years of graduating, she found herself among the first women to drill in the Mediterranean and the lone woman on a rig with 179 men. It is her belief that women don't join the industry, not because of the challenges – which she relishes – but because of the lack of female role models, which is one of the reasons she is willing to put herself in the public eye (Hegab, 2017).

Anne-Christine Dreue

Anne-Christine Dreue, Vice President Business Development EMEA/Fokker Aerostructures B.V. – the aerospace and defense industries also being male domains – speaks five languages, holds a commercial pilot's license, and earned her Master of Science (MSc) in Mining / Petroleum Technology at the Norwegian University of Science and Technology. In 1981, she became one of the first female drilling assistants in the North Sea for Elf International/Elf Norway, calling it, "A very special time for the offshore industry and the position of women in this male-dominated world" (Dreue, 2016).

Melissa Clare

Entering Robert Gordon University in Aberdeen in 1994, Melissa Clare had not yet decided her career path. "I chose to do engineering as it would leave lots of doors open, lots of different applications, and different industries to go into. Living in the oil capital of Europe, I took the opportunity to study mechanical and offshore engineering, in case I wanted to pursue opportunities in the oil and gas industry. Initially there were three females on my university course but, within a few months, I was the only one remaining."

She joined GlobalSantaFe's well engineering development program in 1998 as one of six trainees (four men and two women), spending her first five years offshore. In 2005, at the age of 30, Melissa was promoted to rig manager, making her the first female rig manager in the UK and one of the few in the industry.

Eighteen years after first setting foot offshore as a roustabout and having a successful career with major drilling contractor, Transocean, Melissa felt it was time to leverage her skill set and boldly made the decision to work as a consultant. "I established Petromac (based on my initials) and was quickly invited to support the executive team of start-up company, Borr Drilling, during its rapid growth phase." Having held residential management positions in Canada, Nigeria, Switzerland, and Malaysia, she was well-equipped to provide strategic, commercial, and technical advice. "While it was both fun and rewarding, my longer-term plan was to move to the operator side of the business. I achieved that objective recently, joining Hurricane Energy, where I now lead the Aberdeen office as general manager."

Having joined Lean In Energy[9] as a mentor, Melissa is passionate about supporting women in traditionally male-dominated work places. "It's come full circle, from having been mentored to now mentoring."

She calls the effort to bring more women into the industry, "A movement, a mission, and a journey," and has a philosophical view toward achieving a more gender-balanced workforce. "The situation won't change overnight, but every effort contributes towards swinging the pendulum. I'm definitely a champion for equality and I think we're advancing all the time. But I think that we will only have succeeded when we don't differentiate and we no longer hear statements such as, 'She's the first,' or 'She's an ambassador for women'" (Clare, 2019).

Vicki Hollub

Founded in 1920, Occidental Petroleum Corp. ("Oxy") would not appoint its first female CEO until 2016 when Vicki Hollub, a mineral engineer and 34-year veteran of the company, took the helm. When asked about her history-making achievement, she defers to other women. "There were the first women to go out and work on platforms in the Gulf of Mexico and the first women decades ago to go out and work in the rough, scary North Sea. Those were true trailblazers" (Blum, 2018).

As of this writing, there are several prominent women that reached the upper echelons of the industry and made a lasting impact, but currently are embroiled in controversy. It is unfortunate that these allegations, should they prove to be true, will tarnish not only their reputations, but their contributions to the industry. Among them, they achieved many firsts.

Nigeria's first female Minister of Petroleum Resources (2010 – 2015), Diezani Alison-Madueke, received a bachelor's degree in architecture from Howard University in the US (Akinbajo, 2015) and an MBA from the University of Cambridge in the UK. A Shell employee earlier in her career, she became the first female president of OPEC upon her election in November 2014 (effective January 1, 2015).

Karen Agustiawan, a physics engineer whose career included many years with Mobil Oil, was named president and CEO of Pertamina, Indonesia's national oil company in 2009, becoming the first woman to

---

[9] An independent, non-profit organization affiliated with LeanIn.org. Source: Lean In Energy

be appointed director of a state-owned company in that country. In 2014, *Fortune* magazine named her among its top 50 most powerful women. That same year, she resigned in order to spend more time with her family and to accept an offer to lecture at Harvard University.

Maria das Gracas Silva Foster, who holds a bachelor's degree in chemical engineering, a Master in Nuclear Engineering, and an MBA, interned at Brazil's state-owned oil company, Petrobras, in 1978. Thirty-four years later, she would become not only the first female CEO of Petrobras, but the first woman in the world to head a major oil and gas company, making *Time* magazine's list of 100 "icons who are defining the world in 2012." Once known as "The Iron Lady of Oil," she stepped down in 2015. The following year, it was reported that she is studying for a law degree, which she is due to complete in 2020 (Viegas, 2016).

At the same time, there are women who are breaking the gas ceiling and opening doors for other women to enter the industry every day. They are pioneers who have changed the course of history in this industry. Here, in their own words, as told to the author, are their stories.

♀ ♀ ♀

2	Margaret McMillan
	Water Safety Pioneer

It was an era *"when a woman working her way
into a man's field was just unbelievable."*

In all likelihood, the offshore safety and survival training courses that the women in this book, and the thousands of other men and women working offshore, have taken (and continue to take to keep their certifications current), were created – or at least influenced – by American water safety pioneer Margaret Mary McMillan.

Born August 9, 1920, in the small town of Gramercy, Louisiana, in the American South, Margaret was a water baby, learning to swim at the age of two or three. She told a reporter with the local *Daily Advertiser* newspaper in 2006, "I don't ever remember being afraid of water in my life. It was fun and games" (Guidry, 2016). The rest of her life would revolve around what she referred to as "aquatics." She joined a swim team at the age of eight and later became a competitive swimmer.

Although she never tried out for the Olympics – one of the ultimate goals for a competitive athlete – Margaret eventually would receive an Olympic gold medal at the age of 85. One of her swim students, Dave McAllister, who had been turned away by a number of facilities because he has Down syndrome, had gone on to compete in swimming in the international Special Olympics in 1983 and again in 1991. His mother, Glenda, says, "She was a remarkable woman and she gave him that life skill" (2015). At a celebration of Margaret's career in 2006, in a show of appreciation for her influence in his life, Dave presented "Miss Mac" with one of the gold medals he had earned at a state Special Olympics swim meet.

Margaret McMillan (center)

Tall with strong features – dark, heavy eyebrows reminiscent of American film star Joan Crawford – and a regal bearing, Margaret McMillan was known to strike fear in the heart of many a man, but during her 75-year career as a swimming teacher and water safety instructor, her own heart held a special place for children.

"Margaret, a magnificent mentor of mine, and I were very close," says longtime friend, Bonnie Maillet, CEO of Boysenblue/Celtec, Inc., an oilfield products company in Lafayette, Louisiana, as she recounts another story that illustrates Margaret's devotion to her students. Margaret had shown up to attend Bonnie's 50th birthday party being held at a local restaurant, but had to leave early to teach her swim classes. Bonnie noticed Margaret's "perfectly manicured nails" were painted bright red with dots and clowns on each nail (long before such embellishments were fashionable). Wiggling her fingers, she explained to Bonnie that her nails were intended to distract the children who were afraid of water (2015).

A 1940 graduate of Southwestern Louisiana Institute (later, the University of Southwestern Louisiana, and now the University of Louisiana at Lafayette),[10] Margaret earned a bachelor's degree in physical education and a master's degree in health, physical education, and psychology from the University of Texas. She did post-graduate work at both the University of Southern California and the American University in Washington, D.C. (no byline, 1977).

---

[10] From 1898 to 1921 – Southwestern Louisiana Industrial Institute (SLII); 1921 to 1960 – Southwestern Louisiana Institute (SLI); 1960 to 1999 – University of Southwestern Louisiana (USL); 1999 to present – University of Louisiana at Lafayette (ULL). Source: University of Louisiana at Lafayette

While Margaret has been a source of inspiration to many, the American Red Cross played an influential role in her life. From an early age, she participated in its training programs, and later used her certifications to become an instructor and trainer. She showed her allegiance to the organization by joining the Red Cross during World War II. In conversations with his aunt in her later years, her nephew, Wikoff McMillan, was able to glean some interesting details about that time in her life. "She was stationed down in Miami, probably around 1945 or '46. Her job was to greet the boys coming back from war. She and a few other girls were the first women the soldiers saw in a long time. She had many proposals for marriage but turned all of them down!" (2015).

In 1951, a year after the start of the Korean War, Margaret took a leave of absence from her position as assistant professor of health and physical education at the University of Southwestern Louisiana (USL) after being appointed a general field representative for the organization (no byline, 1951).

A year later, she returned to USL, where she would continue her career, immersed not just in swimming, but all things related to collegiate athletics and numerous other campus activities. During her student days at SLI, she was vice-president of the Red Jackets service organization, which organized and directed half-time shows at basketball and football games, and would later serve as the Red Jackets' faculty advisor. With the inclusion of her own school years, she would devote over 40 years of her life to her alma mater.

Her retirement from USL was announced in December 1977 and just a few days later, Mayor Kenny Bowen proclaimed December 6, 1977, "Margaret McMillan Day" in Lafayette, and a celebration was held in her honor to show the city's appreciation for her contributions.

It would be almost impossible to overstate Margaret's standing in Lafayette, whether people knew her from her days as a swimming instructor or as an international water safety expert. Dave Domingue, operations coordinator at Lafayette International Center,[11] says, "Margaret was quite an amazing woman, and I – like so many others – knew her from my childhood swimming lessons in the early '60s. She loved aquatics and wanted to share that with everybody. She was always very dear to me. She made an obvious and lasting impact on an industry that is integral to our local economy, and I'm pleased that her

---

[11] A division of Lafayette Consolidated Government

accomplishments will be given additional acknowledgment. This woman touched everybody in Lafayette at the time" (2018).

Three days later, the celebratory mood ended with the news of an offshore helicopter accident in the Gulf of Mexico on December 9, 1977, that took the lives of 17 of the 19 men onboard, almost all of whom were young men in their 20s and 30s. The accident weighed heavily on Margaret's mind.

Many years later, in an interview, she would recall thinking to herself at the time, "My gosh, why can't something be done about this?" (Pratt, 2004). She was unable to find an underwater egress training program anywhere in the United States. Even prior to the accident, Margaret had been dismayed to learn that although companies were required to have safety equipment, they were not required to train the (mostly) men how to use it.

Because of her reputation in aquatics, she had been approached about offering water safety training to petroleum company employees – as well as various branches of the US military – and had been doing so for several years before taking early retirement from USL. The courses received an enthusiastic response from the men, many of whom had served in WWII and remembered warfare aquatics, and word spread. She said it was "one of the most exciting experiences of my life" when a group of oilmen that had taken her safety training course stood up and gave her a five-minute standing ovation (Pratt, 2004).

Margaret, who traveled the world and worked alongside men of international renown, claims rarely to have encountered what she calls a "male chauvinist," although she distinctly remembers one man, whom she says pointedly was "from the United States of America," who was "infuriated" that a woman was representing the US at the 1977 Intergovernmental Maritime Consultative Organization (IMCO)[12] meeting. She didn't seem to take it personally and chalked it up to being an era "when, you know, a woman working her way into a man's field was just unbelievable" (Pratt, 2004).

After taking early retirement, Margaret had established her own company, McMillan Offshore Safety Training (MOST), which would go on to become a family business. Several of her nephews, who had grown

---

[12] Known as the Intergovernmental Maritime Consultative Organization (IMCO) until 1982, the International Maritime Organization (IMO) is the United Nations specialized agency with responsibility for the safety and security of shipping and the prevention of marine and atmospheric pollution by ships. Source: IMO

up taking swimming lessons from her and who now shared her passion for safety, joined the new venture.

"I do not know how to say this without getting a little emotional," she said in a 2004 interview, "because I did not do it by myself. Nobody ever does anything by themselves. And sometimes the others who have helped and contributed do not get the credit. There were so many people that were very instrumental in this total thing. One of the things that I am particularly proud of is the fact that I have four nephews, three of whom – and this is part of my legacy – have followed me into this world" (Pratt, 2004).

Margaret would make numerous trips overseas, gleaning as much as she could from other countries, such as England, Scotland, and Norway, whose water survival training programs were more advanced than those in the US. Her travels abroad also included attending nine meetings of the IMCO during the time she was associated with the US Coast Guard. The first one was all the more memorable because it was then that she met Dr. Joe Cross, the managing director of the Robert Gordon Institute of Technology (RGIT) Survival Centre in Aberdeen, a legend in offshore survival training, whom Margaret considered her mentor. She also was delighted to learn the Coast Guard had called her its "secret weapon"! (Pratt, 2004).

In 1988, Margaret was instrumental in the creation of the Marine Survival Training Center at the University of Louisiana Lafayette. At the time, it was the only facility of its kind in the United States. It has grown to become world-renowned, with a state-of-the-art modular egress training simulator – a far cry from the days when Margaret trained men to escape from a straight-back chair submerged in water!

In 2004, Margaret became the first woman to be inducted into the Offshore Energy Center's Hall of Fame in Houston, Texas, where her contributions to offshore safety and survival training were recognized under the Health, Safety & Environment category in Pioneering Technologies.

While she is in illustrious company – Paul N. "Red" Adair, former US President George H.W. Bush, and Billy Pugh are among her fellow Hall of Fame recipients under various categories – she remained the only female inductee until 2016 when fellow American Lillian Espinoza-Gala was inducted as an Industry Champion for her work in Health and Safety. (Each year since then, at least one woman has been inducted.)[13]

In a wide-ranging interview with Dr. Joseph (Joe) A. Pratt, Professor Emeritus at the University of Houston and a leading historian of the petroleum industry, prior to her induction, Margaret shared some fascinating insights about how offshore safety evolved from the old "run and jump" strategy, making sure to give credit to the many people – men – who had inspired her along the way and shared her mission to make the industry safer for everyone.

The woman who traveled the globe, mingled with heads of state, and was the recipient of numerous awards, later said, "Being recognized as a leader by [my] peers and being the first woman to get this honor was one of the most exciting highlights of my life" (Landry, 2004).

While Margaret never married or had children of her own, her brother's five children – four sons and a daughter under the age of nine – became her children, in a sense, with his untimely death at the age of 42. Margaret helped his widow, Rosa McMillan, raise the children and was such an influence in their lives that three of the four boys – John, Wikoff, and Haas – became involved in water safety in one form or another. (Their brother James did work as an offshore surveyor for a time although, ultimately, he and their sister Robin – who is in the hospitality industry – chose unrelated careers.)

In the early days of the oil and gas training, the McMillans had to convince companies that their program was worthwhile as there were no standards in place at the time that would require companies to offer such training. John, III, Margaret's eldest nephew, says because of his aunt's high safety standards, "We always believed that when a man ventures out to sea, his life is in jeopardy."

Week in and week out, they would make the 400-mile round trip from Lafayette to Houston, stopping in various coastal communities in Louisiana, Mississippi, and Alabama to provide training.

"I once told Aunt Marg in a voice of anger that this lifestyle was getting old. She looked at me and said, 'If you keep doing what you're doing, you will get your name in more places than just the phonebook.' To this day, I laugh at that comment. It calmed my nerves and I continued to travel... and travel we did" (2019).

---

[13] In 2017, Anne Grete Ellingsen was inducted as an Industry Champion and in 2018 both Sara Akbar and Eve Howell were inducted as Industry Champions.

John attended safety conferences around the world with Margaret, traveling to training centers in Scotland, Norway, Australia, and Denmark and is now the US representative to the International Association for Safety and Survival Training (IAAST), an organization his aunt was instrumental in forming.

As much as he loved and respected his "Aunt Marg," as he and his siblings called her, she cast a long shadow and he wanted to achieve success on his own terms. While staying involved in water safety, he chose to focus on the maritime and fishing industries. Based in Maine, where he teaches Ocean Survival at the Maine Maritime Academy, he carries on Margaret's legacy through the name of the family business, McMillan Offshore Safety Training (MOST), which Margaret established in Louisiana in the late '70s.

Wikoff, second in the line of brothers, stayed close to home and is safety director at Aries Marine Corporation in Lafayette, where it seems almost everyone took swimming lessons from Margaret. "In my younger days, I attended a few aquatic schools and always met someone who knew Margaret or was taught by her. She was one tough swimming instructor. I hear many stories about how "Miss Mac" did not give any slack." He is quick to say she had her students' best interest at heart because she knew they could do better. "In the end, those that stuck with her appreciated what she did for them. She was very good at what she did and, when she committed to something, it was all or nothing" (2019).

That dedication and determination not only served Margaret well throughout her life and career but her nephews agree it made them the men they are today.

After more than 31 years with MOST, the family business started by his aunt, the youngest of the brothers, David (who goes by his middle name Haas), left to become involved with other water safety and survival training programs. With Shell since 2008, he now oversees Training & Operations for the Marine Survival Program at Shell's training center in Robert, Louisiana. His Aunt Marg's presence is never far away; a curio in the survival center displays her major achievements and awards in dedication to the contributions she made to the industry.

Haas relays a conversation that took place between his brother John, Terry Crownover, who worked for MOST and is now the Director of MSTC,[14] and himself at an international survival conference in Halifax.

---

[14] The University of Louisiana Lafayette Marine Survival Training Center

"John was talking to me and Terry, who was considered the adopted fifth McMillan brother and part of Aunt Marg's legacy. Looking around the room with all the world leaders in training, he said, 'Look at Marg's legacy. You and Terry run the two world-class training centers in the US.' I then said, 'Don't count yourself out; you are right up there with us.' It really put into prospective where I was in life.

"We grew up in this; it's a way of life. Sometimes, it takes someone else to point out to me the importance of what I'm doing."

He never forgets the contributions his aunt Marg (which sounds like "Mawg" in his Cajun accent) made to the industry nor has he forgotten the people who helped her achieve her goals.

"My gawd, think about it. A woman in this business in the '70s?" He believes her success was due, in part, to several key people (men), including Johnny Ryan, whom he calls "a true American success story. He started out packing life rafts for Mr. Alexander at Alexander Industries and ended up buying the company! He's an old trooper, like Aunt Marg" (2019).

A bronze plaque engraved with Margaret's favorite expression regarding safety hangs next to the curio at the survival center:

> "If to you life is sweet,
> use your head, and not your feet."

It was a lighthearted way to convey a serious message: the old safety advice, "run and jump," could be deadly.

Now 95, Margaret McMillan resides in an assisted living facility in Lafayette, Louisiana. Due to health issues, she no longer gives interviews. Her friend, Bonnie Maillet, boldly proclaims, "Margaret is responsible for saving more lives on water than any other person on this earth." However, her nephew Wikoff McMillan says, "Margaret is a living legend and wants no credit."

♀ ♀ ♀

Margaret McMillan (August 9, 1920 – August 31, 2016) passed away during the writing of this book at the age of 96.

The Margaret McMillan Papers are housed at the University of Louisiana at Lafayette. Collection 289. McMillan, Margaret (1920-2016). Papers, 1909-2016.

Yassmin Abdel-Magied
Mechanical Engineer

"I never walk into a room thinking I have no voice;
I tell myself, 'I belong here.'"

Yassmin Abdel-Magied once had a mentor who said, "I know you're all the things that don't belong on an Australian rig, give it a go anyway." Fortunately for the oil and gas industry she did because, ultimately, it was not a question of whether she was right for the industry but rather whether it was the right industry for her.

Born in Khartoum, the capital of Sudan, Yassmin, whose family is a "mix of Sudanese and Egyptian with a sprinkling of Turkish and Moroccan," is the daughter of an electrical engineer father and an architect mother. She describes her father, a then-23-year old PhD holder from Imperial College in London, as "incredible" and the "star of the faculty" at the University of Khartoum where he was a lecturer. However, that status did not protect him – or his young family – from the political instability of the early '90s. When a new government took control, her father protested a move to change the language of the university from English to Arabic and, as a result, lost his job. Given the uncertainty of the future, he, his wife, and baby daughter immigrated to Queensland, Australia, where they eventually obtained dual-citizenship, and a few years after settling, Yassmin's brother, Yasseen, was born.

Raised by her parents in the Muslim faith, Yassmin attended a Muslim primary school on the south side of Brisbane in Queensland, where she began wearing the *hijab* – or headscarf – when she was about 10 years old. "It was something I wanted to do," and although her mother does wear the *hijab*, Yassmin says her parents didn't encourage or discourage her either way. "They just wanted to make sure it was my choice."

She later attended an ecumenical Christian high school. "I guess my parents wanted to give me a bit of variety, give me the best education they could." The school had a student body of about 2,500 and Yassmin was the first Muslim girl to wear the *hijab*. "I definitely was conspicuous but I was somebody that did things the way I felt was right, and I did what I wanted to do regardless of whether or not that's what everybody else did.

"I'm very fortunate that my family played a big role in engendering that. Growing up, my parents never praised me on what I looked like. My value was never tied to my visual or physical appearance; my value was always tied to my outcomes and my output and my behavior, and that's how I learned to judge who I was as a person."

Despite wanting to remain in their chosen fields, both of her parents had to leave their professions once the family emigrated from Sudan due to a lack of opportunity, but also the refusal of formal recognition of their degrees in Australia. Even so, family conversations were technical – and political.

"We always grew up fixing things and thinking about problems and designing solutions. It also meant there was never any gender associated with technical fields. My dad would never say something like, 'Men do engineering,' particularly because in Sudan architecture was considered part of engineering so, in my mom's mind, she was an engineer. I didn't think about engineering or architecture as a gendered profession. It was only when I started becoming aware of what society said about these professions that I began to think of them in any gendered manner." Because both of her parents are technical, Yassmin believes it gave her a balanced view of the professions themselves as opposed to what they represent in society.

She describes her dad as "very supportive, very encouraging" and says he would help her with math and science, particularly in high school, and was always interested in teaching his children. "When the tap was leaky, he would say, 'Okay, kids, we're going to change the tap, we're going to go build this.' He always got me to come along. There was never any difference between me and my brother in learning."

While growing up in a technical environment was part of her everyday life, it was seeing a movie at the age of 13 featuring a young boy who did "go-karting," that made her fall in love with the world of motorsport and soon after she realized, "Engineering would be such a tangible way of being able to help people and being able to contribute to society, so the two worlds and two passions really linked."

Although Yassmin had a desire to help others from a young age, she experienced a life-changing event at the age of 16 that motivated her to act. As one of 100 young people attending the Asia-Pacific City Summit, she heard about various organizations that were working on similar concepts to create awareness and combat social issues such as youth homelessness and domestic violence. As exciting as the atmosphere was, she noticed the lack of cooperation among groups, as well as the competitive nature of the conversations. She thought, "Surely, we are beyond this. We should be focusing on the people and not on who gets the credit."

At home that night she shared her frustrations with her mum, who responded, "If you think something's wrong, why don't you do something about it?"

Yassmin remembers thinking, "Touché, Mother, touché."

From that challenge, Youth Without Borders (YWB) was born. Initially, the idea focused on getting young people to collaborate to implement positive change in their communities. Naturally, she was met with skepticism and told, "You're 16 years old. Do you know how hard it's going to be?"

"And I didn't. But I did it anyway," she says. "I think, partly, my youth and my naiveté helped me because I didn't know what I was getting myself into. I went into it blindly. I was like, 'This is what I want to do. Watch me do it.'"

"We try to empower young people as leaders of positive change through building their capacity and also allowing them to run their own community and collaborative initiatives. Essentially, young people come to us with an idea and say, 'I want to help my community this way,' and we try to help them make it happen."

The organization's first initiative was called *Kamar Buku*[15] and brought together nine different groups in Indonesia to set up mobile

---

[15] Loosely translated as "the Book Room." Source: Google Translate

libraries, which transport boxes of books on the backs of motorbikes from town to town. The project continues to run and is owned by a group in Indonesia, which Yassmin says, "For us, is the most important thing."

Her favorite project, not surprisingly, is an engineering camp for students in grades 10 – 12 who face some sort of barrier toward further education – whether they are the first to attend university in the family, from a disadvantaged socio-economic background, are Indigenous or may be asylum seekers – and who do not have the means to afford such a camp.

"We recognize that in Australia, there are a lot of opportunities but not everybody has fair access to them. When we select these students, it's all about showing them that anything is possible, and it's possible for *them*, and showing them how to do it. We want to provide this life-changing experience without the financial burden getting in the way. It's often educational, sports-based, and cross-cultural but all of it is about, How do you implement positive change? How do you help people get access to the opportunities, so that they can realize their own potential?"

While helping others unlock their potential, Yassmin was still exploring her own. In high school, she pursued non-traditional interests, winning the bench press competition – "that was pretty cool" – and was the only girl in the senior school of design and technology wood-working class. After she graduated from high school, she also trained as a boxer for five years competing in amateur spars (although her mother didn't know she fought!).

Graduating with an Overall Placement (OP) 1, the highest possible mark in the state, she enrolled as a mechanical engineering major at the University of Queensland, where she began pursuing her interest in race cars in earnest, becoming a race car engineer, designing her own race car chassis, and running her university's race team.

After graduating as valedictorian of the Class of 2011 with a Bachelor of Mechanical Engineering (First Class Honors), Yassmin enrolled in a very selective Master of Motorsports in the UK, which she describes as "a shoo-in to an F1 job." She was offered work experience at Mercedes. She had found her career path working in Formula One. "I'd made it, you know. I had hustled my way, somehow, someway, into being able to make it into the F1 world."

Then two things happened that would change the course of her future. She did not receive the scholarship for the motorsports master's that she had expected to get, which left her unable to afford the program. The second was, upon returning home from her first trip to the UK to visit different race teams, she had a "niggling bit of self-doubt" about whether or not she wanted that to be her career path.

"The reality didn't quite match expectation," she says, "but I wasn't really ready to admit that yet, like, 'Maybe it's just a feeling I've got, I'm just going to continue.'"

Forging ahead, she needed to find a job in order to save $50,000 in a year to be able to enroll in the master's program. She thought about working in a field that would enable her to do so and recalled a woman in the booth of a major service company she had spoken to at a careers fair in the first year of engineering. At the time she had thought that sounded "cool," but had done nothing about it.

A couple of years later in third year, she participated in a mock interview designed to help engineering students get a job. Ironically, the interviewer was with the same major service company. Yassmin was his last interview for the day, and he told her she was the first person that day to pronounce the company name correctly. That must not have been the only thing that impressed him because, when the interview was over, the man told her if she ever needed a job, to e-mail him.

"It was just a whim, but I was like, 'You know what, I'm going to e-mail that guy and ask him if he still has a job available.' And he did!"

The company then offered Yassmin a field engineering position and she accepted but, in her mind, it was only a means to an end. At that moment, she decided she would take the job for a year, save enough money for the motorsports master's program, and then pursue her dream of a career in Formula One. In a conversation with her mentor, she said emphatically, "I am going to do this for a year – *a year only* – because this is not the career path I'm going to choose." He said, 'Everybody says that,' and I was like, '*No!*' because I wanted to work in motorsports."

She laughs at the memory, "The rest, as they say, is history!"

For the next two years, she did field work for the service company as an on-call measurement while drilling (MWD) engineer.

"When I got hired in 2012, I was in this particular segment that had never had a female field engineer before in Australia. I made sure that I referred women, and three of the next six hires we had were women. It was an interesting time. I was lucky in that the first rig I ever went to there was a female geologist. But, in my year and a half, two years, in the field as an MWD, I came across very few female engineers and field technicians, and no women on the rig crews themselves. Bear in mind, I was on land. Offshore there are many more women than there are on land. By "many more women," I mean there might be one or two other women on the rig," she says, laughing. "The percentage doesn't actually increase but you feel like there's more of ya! To work with another female is sometimes just refreshing."

She then jumped ship for a drilling engineering position with one of the majors. "I did pretty well, really," she muses. "This job was the jackpot!"

Again, she remembers feeling very indecisive about whether or not she would take the role because about a year and a half into the job with the service company, another mentor had asked, "Yassmin, what are you doing?"

"Because I was a field specialist, I wasn't really using my mind for what it had been trained for – engineering and design. It also seemed that I was only seen as an operator. Although there is nothing inherently wrong with that and I really valued the experience, I just found I wasn't able to bring my whole self to work. I had first-class honors, I was running a youth organization on the side, doing all these speaking engagements, but none of that really mattered to my job, and I just had to accept that – or move on." Her boss at the time wasn't interested in her life outside work.

"He was like, 'Look. When I want you to be on the rig, you need to be on the rig.' From his point of view and the company's, I was an *asset,* and from my point of view, I was trying to manage my life. It was becoming more incompatible as time went on, even though I initially had thought it wouldn't be an issue."

It was then that her mentor suggested she find a company where she would be happy but also one that she would be invested in.

"It was actually really difficult for me to make the switch from being a service hand to being a company person. Being a company person is an entirely different mentality. It's one I'm still getting used to," Yassmin admits. "It's like being in the trenches versus being someone guiding the troops. A difficult transition, particularly if you live for the banter in the trenches!

"As a service hand, one of the ways I would get credibility with the rig crew was having the best graffitied hard hat. I quite love a bit of ink art, so I always used to deck out my hard hat in really awesome graffiti. That was my pride and joy. Across Oz, there are guys walking around with my tags all over their hard hats, which is awesome! It's so much fun. As soon as I became a drilling engineer, the Health Safety and

Environment (HSE) guy said, 'You can't walk around with that hard hat.' I was like, 'What do you mean? This is a great hard hat.'

'Absolutely not. You're the company now, and you have to be respectable, and follow the rules.'

"Unfortunately, gritting my teeth, I had to retire it. It's in my bedroom now."

Yassmin reiterates that she genuinely did not start out wanting to be an oil and gas engineer but discovered she really enjoyed working in the field and that is what led her to stay. In her current job, she divides her time between the field and the office. She admits if she had started in that role, she is not sure she would have stayed on.

"I think there was something about the freedom of being a field engineer that was just really liberating," she says. "When you were on the rig, you just needed to do your job and then everyone else would leave you in peace, particularly the role I was in. I remember asking a company man once, 'What makes a good MWD?' and he said [imitating a man's voice], 'When I never have to hear from ya 'cuz if I'm hearing from ya, it means something's gone wrong.'"

And go wrong it did. On the "well from hell," Yassmin was the lead MWD, and her relief, who worked nights, was on his first rig. The job involved a complicated design and Yassmin found herself working way too many hours a day (for which she was reprimanded) and getting called back to the rig when she tried to go to the base camp.

"I was sleep deprived and it was the fourth week I had been on this job. I remember sitting in my room, going, 'What am I doing? Why am I here? What is happening?'" Ever the optimist, she says, "But then you make it through and it's a great story."

It is those types of experiences, Yassmin says, that teach you to depend on yourself and to believe in your ability to cope with difficult situations. Once you do, the reward is greater self-assurance and confidence in your abilities to handle the next challenge, which can occur on a daily basis offshore.

"Every day is an adventure. And, yes, it's a politically sensitive industry. And, yes, it can be tough. And, yes, as a young woman you have to work twice as hard to get the respect. And you have to deal with older generations of men who may wonder if you have earned your place. But how rewarding is that challenge? Where else are you going to get that type of experience? Where else are you going to be able to spend time with people from such different worlds?

"We always joke offshore, we're all crazy to be out here, we're all crazy in some way, but that craziness is what makes it interesting and what makes it exciting."

It is those shared difficulties – and victories – that create the bonding experiences offshore workers talk about so often. "I definitely know people that the environment has not worked well for, but then they just leave. I guess the caveat is, as long as you can handle the environment – that's not necessarily a gender thing – it's the isolation, it's the routine way of life, it's the way people speak to each other – and are okay with it, then it can be a really enjoyable experience. But if that is something that does not work for you, then everything becomes infinitely worse for men and women, both, in equal measure."

Even though she has seen "military blokes," men who have done tours in Iraq and Afghanistan, not be able to complete a hitch on the rig – with one young man even having a breakdown and having to be taken off the rig when she was a service hand – Yassmin says when she asks the men she works with whether they would "let" their daughters or wives work in the industry, "Almost all of them say, no. Not even 'no,' they're like, 'No, never.'"

She has found that male engineers are open to their wives or girlfriends or daughters working in the industry, but that male oil and gas workers often are not. "One of them said to me, 'Because I know what it's like out here.' I think that's the difference. But, as with anything, there is a range of opinions. Some guys are supportive, even encouraging. Things are changing with each generation."

She also believes a girl's father often plays a role in whether or not she will consider a career path in the industry. "Daughters may be given all the positive influence in the world but if their father says no, then it's much more unlikely she will choose a path that her father doesn't support, particularly if she's somebody that looks up to her father."

Despite being generally supportive of his daughter's aspirations, her own father wasn't impressed when she accepted the position as an offshore drilling engineer because he couldn't understand why his daughter wanted to work "in the middle of the ocean over a job in the city where we lived. It just did not make sense to him. And, from a cultural point of view, it was not okay. However, I have always been of the opinion, as long as I am adhering to my religious values in a way that I deem has integrity and I communicate that to the people around me – verbally but also through my behavior – then it's usually fine."

Yassmin says she has been surprised to find that people are more unsure how to deal with her gender than they are with her faith or her color, and her expectation that her faith would be a bigger issue has largely proven not to be the case.

"The thing is, the guys I work with don't realize that I'm Muslim, usually – not off the bat, not unless I tell them. It's hilarious. I will say to them, 'Guys, can't you tell that I am Muslim because of this thing I wear on my head (the headscarf)?' I've gotten all sorts of responses, from, 'We thought you had dreadlocks' to 'We thought it was an Occupational Health & Safety (OH&S) thing.' It's fascinating because it means they get to know me as a person first before they apply the opinions and stereotypes of Muslims that they would have had."

She recounts a story that illustrates the power of unconscious bias, one of her favorite topics of conversation, whether she is "hanging out with the boys occasionally at the smoko shack on the rig," or giving a TED talk before hundreds of people. She came across a male worker on the rig, reading a motorbike magazine. Being the motorsports enthusiast that she is, she approached him and asked if he rode. Their common interest led them to strike up a conversation about motorbikes that lasted for the next several weeks while they were on the rig. A few months later, Yassmin received a message from a friend who asked, "Do you know a guy named John?"[16] She responded that she did and asked why.

Yassmin's friend said she knew John's sister and his sister had told her a *weird* story about how John, who previously had held some closed-minded views, had suddenly become a staunch defender of Muslims, telling people that they shouldn't disrespect others. Her friend concluded, "Apparently, it's because he met some girl named Yassmin on a rig. What did you do to him?!"

Laughing, Yassmin says, "I love that story because it illustrates the power of human connection over any type of bias. For these guys, I am often the first Muslim woman they've ever spoken to and that connection combats the negative stereotypes they may believe."

Yassmin actually sees working in the offshore industry as a way to facilitate her social activism. "I can definitely imagine a future where I work a month on/month off on a drilling rig for the rest of my life and, in my month off, I do whatever I want, which includes running youth

---

[16] Not his real name.

organizations, agitating for change, talking about unconscious bias, and running workshops," she says, enumerating a list of her passions.

"The reality is, it's still incredibly difficult for a woman to make it through the offshore world once you start having different priorities, like family. As a single woman, it's wonderful. I travel all around the world, I do whatever I want, and it's great. That being said, it's incredibly tough on relationships. So many men I've worked with are divorced, so I have realized it can be difficult to have a family with this lifestyle. I also think I will find it tough to find a bloke that is okay with that kind of lifestyle for his partner. Time will tell!"

There is a lot of talk within the industry about attracting female talent and Yassmin gives her current employer credit for making a concerted effort to hire women. "In my first year, they hired three female drilling engineers and one male, which is the most women I've ever worked with in my life." However, she sees retention as just as great an issue, for herself as well as other women. As much as she loves the industry, she feels a certain amount of ambivalence regarding the longevity of her career.

"As long as industry stays relevant and stays responsive, and not defensive and protectionist, in the face of great change, then it will be a good place to be a part of. But I want to be a part of an industry that tackles energy problems head on, that realizes that we live in a world that is not just about profits but also other facets, so it will be interesting to see what the next few years bring.

"The trick will be, how do you get women to stay? I think there's the element of getting girls to see engineering and Science Technology Engineering and Mathematics (STEM) subjects as an option, getting chicks to choose oil and gas as an option, but then how do you convince them to stay because the environment is still not conducive to all the 'life stuff,' like having a family."

She believes part of the solution lies in mentoring and sponsoring and says she owes a lot to the people who have been willing to put themselves on the line for her and put her name forward when the occasions have arisen.

"They have definitely significantly shaped my opportunities. I always encourage people to find other people they connect with, and also not be afraid to ask because people can't read your mind."

She speaks from experience, having recently been the beneficiary of a high-powered mentor. This supremely confident young woman admits she was scared; she wanted to make a good impression. A friend advised

Yassmin to be upfront. "Tell him straight out what you want. How is he going to know you want this position, if you never, ever actually said you're interested? He's not going to know."

"Fair enough," Yassmin said. "I thought that he would guess."

Her friend responded, "He's too busy to guess. That's not his job to guess what you want. Tell him what you want and then work together to make it happen."

Yassmin reiterates, "People can't read your mind. It's common sense."

As women move into leadership roles, they inspire other women to reach higher. "Even in places like Sudan, that have all sorts of other struggles, there are places for women in positions of authority. I love being able to see that side of the world to know that the story that's told about people like me is not necessarily true.

"We can't ever underestimate the power of role models. The first time I met a senior female drilling engineer, I almost lost my mind! I was like, 'Another real chick drilling engineer exists, and one that has been around for a while to boot?!' Every single person feels like they're forging this new path, but when you see somebody ahead of you, and you can talk about the experiences, and they *get* it – oh, there is nothing quite like it."

This is particularly valuable for women who rarely encounter other women in their line of work. While Yassmin has met the occasional woman on an offshore rig, she has never seen a senior Muslim woman on an offshore rig in Australia. "Yes, in Malaysia, Indonesia, Brunei, but those are different cultures. The majority of the time, I work in Western cultures, where I am the only woman of color, so you're tackling a bunch of other marginalized identities. Seeing someone ahead of you who is going through that experience is how you visualize the possibility of being in that role.

"The fact is, I am very young. I started out in the industry when I was 20, I'm now 24, and I'm lucky enough to be in supervisory roles on the rig. When you've got guys that are like, 'Darling, I've been doing this since before you were born, since before your *mother* was born,' you've got to figure out how to influence them in different ways while also respecting their knowledge, and understanding that you don't have all the answers, but you're both trying to get something done, and finding the best way to do that."

On occasion, "the blokes," as she likes to call them, have made jokes about her youth at her expense. "When I first started on the rig, the guys found out that I still lived at home. One of the guys said, 'Mate, you've probably still got [expletive] teddy bears on your bed. I'm going to call you Care Bear!' So on that rig I was forever known as Care Bear. It was awful," she says, but her laugh tells you she was in on the joke.

Often being the youngest one in the room – or on a rig – has never intimidated Yassmin and she believes that self-confidence is something her parents gave her. "I never walk into a room thinking I have no voice; I just walk in and consciously tell myself, 'I belong here,' and consciously think about the fact, yes, they have experience and I need to respect that experience. But I also try to remind myself, 'I am here because I belong, too.'"

A male colleague once said to her, "Yassmin, I don't understand why you women are so caught up in how you get opportunities. If there is a quota, why don't you guys just take it? If there was a quota that said because I'm a man, I'm going to get this job, I'd bloody take it! I'm not going to be like [she feigns a man imitating a woman's voice], 'Nah, I'm not going to take it because I want to be given it in a proper manner.' What is it with you people? Just take the opportunities when they come and stop questioning it."

For Yassmin, his comments led to an epiphany of sorts. "It was so fascinating; I had never thought about it that way. We're so wrapped up in *how* we get opportunities when a guy would never question it. So now, whenever one of my girlfriends acts like that, I say, 'Mate, get over yourself. Take the opportunity and then prove you deserve it, not the other way around.'"

♀ ♀ ♀

Since coming onshore, Yassmin has worked full-time as a writer and broadcaster. Her second book, a Young Adult novel called *You Must be Layla*, was published by Penguin in March 2019. She also works with corporations and organizations around the world to tackle unconscious bias and create workplaces and societies that are inclusive, safe, and equitable for all.

4	Sara Akbar
	Co-founder & CEO Kuwait Energy

"When women are given the positions,
they are far more powerful in what they do than men."

Driving into the dense black clouds of smoke, fires raging around her, oil wells still exploding in the distance, Sara felt an overwhelming sense of sadness. "I saw with my eyes the wealth of my country evaporating into thin air." Fires from more than 700 oil wells torched by Saddam Hussein's troops as they fled Kuwait at the end of the 1991 Gulf War had turned day into night and the desert into an inferno.

Help from all over the world, eventually totaling 40,000 people (*Fires of Kuwait*, 1992) – the largest non-military mobilization – poured into the tiny, oil rich nation to extinguish the fires set by Iraqi forces. The world-renowned team of Boots & Coots was there, as was legendary firefighter Red Adair, but Sara could not sit by idly and wait for someone else to rescue her country from certain devastation.

She was a child of the oilfields; they were her birthright. She has said she was born in an oilfield – and, indeed, the hospital where her mother gave birth to her was located in the middle of the oilfield where her father worked for Kuwait Oil Company (KOC). [17] She lived near the oilfields her entire childhood, and would wander through them during the day, as there was less security as there is today, and everything was open. She fondly recalls playing in those fields, close to the oil wells, the fences, even the fires. "That's why it became so familiar and so natural for me to work in these fields."

---

[17] A subsidiary of the state-run Kuwait Petroleum Corporation (KPC).

Despite the fact that her father and mother had no formal education, but had taught themselves to read and write, Sara says both of her parents valued knowledge and education and pushed their children to study – "and study hard," she emphasizes. The daughters, in equal measure as the sons, were encouraged  to earn a degree and "find our own way," something Sara says was not unique to her father. In that era of Kuwait's history, the 1970s and '80s, she says Kuwaitis were far more liberal and accepting of gender equality, and more respectful of women, than they are now. Parents encouraged their children to attend university and earn degrees; it was the norm, and there was "heavy punishment" for not studying and doing well. The discipline was tempered with love, which she calls her mother's "specialty."

"She loved all of us to the point that each of us thought we were her favorite. We never knew, even today none of us knows, who was her true favorite."

It was a powerful combination and all ten children – five sons and five daughters – are university graduates (including two who are PhD holders), something that must have been a source of immense pride to their parents.

Upon graduation in 1981 from Kuwait University with a Bachelor of Science in chemical engineering, Sara became the first woman to work as a petroleum engineer for Kuwait Oil Company. However, in those days, women were not allowed to work in the field and female engineers were hired to work in the office.

Sara says she knew, "If I want to be a good engineer, technically sound, I actually have to have field experience because an engineer with only an office background will not do a good job." She insisted that she be allowed to work in the fields. It was unheard of for a woman to work offshore, but she happened to be in the right place at the right time.

"During that period, we were drilling the first offshore well in Kuwait and I prepared some of those programs to complete those wells and stimulate them. It was just a pure coincidence that those first two years of my work in the company we were focusing on offshore, and that's why I ended up being the first Kuwaiti woman working offshore and onshore."

Sara came to an agreement with the company that she would only work during the day. On her first excursion offshore, transportation problems and treacherous weather conditions delayed her replacement, a male engineer, who later refused to board the rig and went back to land, leaving Sara to work nearly 20 hours over the course of two days including all night.

She couldn't have been happier. Calling the rig "beautiful," she said it felt like home. She continued the pattern of day trips offshore for the next two years and while she found it challenging, she says, "I really and truly enjoyed every single minute. I loved it."

Neither Sara – nor anyone else – could have foreseen how valuable her offshore experience would later prove to be.

She had set a precedent and, even after going onshore, she continued to work long hours in the field, sometimes for days, for the next eight years. Sara says she felt very comfortable in the field. It was an intensely personal experience where, "I felt like myself and that it was where I belonged." Even now, she says, "That was a truly enjoyable experience that I would do again and again and again, if I had to."

Working in the field has its own merits, and seeing the immediate results of her work is something she found gratifying. After implementing the program, she could determine instantly whether her work was correct or not, and if the response of the well or the facility was as she expected. It is actually "engineering in its practice," she explains, an aspect that has great rewards, and something an engineer does not have the benefit of, if she is isolated in an office. Sara says there is also a fast learning curve in the field and engineers acquire new knowledge and skills on an almost daily basis. In addition, she says, it also teaches you that, in the field, there is no difference whether you are senior in age or if you are female – seniority doesn't mean much except in how it pertains to your knowledge, which is what dictates who leads.

"That's how you find yourself in a leadership position. It teaches you how to work in teams, how to command, let other people do what you think is right," she says. "The soft skills are far more obvious and quick in the field than in the office, and that actually leads to a lot of the leadership skills you need in your future."

Her tone is animated when she talks about working in the field and it is apparent that she cannot stress enough the value of practical application in the field.

"The first aspect of working in the field is your technical skills, so you need to get your hands dirty first. That's the number one priority for any engineer." Leadership, Sara believes, should be the second priority, and as you progress in your career, those skills are acquired.

"You need to know how people work in the field, you need to know the quality of the data you receive, how it was obtained, what the margins of error are, so it gives you a bigger picture kind of environment when you become a leader. To emphasize, the first skill that you get in the field is technical ability, so you know exactly what each number means, where it came from, how it was measured, what the accuracy of that number is – very important aspects to any business."

"That's why I encourage women to be in the field – whatever field – get your technical ability right by doing it yourself first, learning it, and then you can sit in an office as an executive."

It was those years of field experience that gave Sara intimate knowledge of both the offshore and onshore wells – "I knew them like the back of my hand" – and would ultimately help her play an integral part in preventing her country's natural resources from going up in flames.

Wild oil well firefighters are an elite fraternity and not only are their numbers relatively few, no firefighting group had ever encountered more than three oil well fires simultaneously in any disaster around the world (Curtiss, 1995). Each well, just like each fire, has its own specific personality, which makes it unique, and knowing the intricacies of the wells is a critical part of being able to successfully bring them under control. Sara had never had any firefighter training, but what she did possess was ten years of field experience. It was that familiarity with the wells that enabled her to participate in the firefighting efforts not just as the sole Kuwaiti woman but as the only woman from any country.

Sara says she and her fellow engineers could not sit and watch the wealth of their country evaporate when they possessed the knowledge that would allow them to do something about it. Realizing this, they banded together to form an ad hoc firefighting team and called themselves the Kuwait Wild Well Killers.

In Sara's assessment, the inexperienced firefighters "did a fantastic job utilizing the knowledge that we had as engineers of the wells and the environment – how do you get water? How do you use water? How do you put [out] the fire? How do you control the wells? How do you rehabilitate the well afterwards?" Still, it was a huge amount of learning to absorb in a very short time under great duress. Sara remembers the intense heat, the asphyxiating smoke – "you can't see much, you can't breathe much," – but says, "We were full of passion and excitement, and that actually made up for the loss of other senses."

At the time, Sara was quoted as saying, "The amount of pleasure and joy I see on the faces of the rest of the team whenever we put a fire out is tremendous. It's like a dream finally coming true" (*Tides of War,* 1992).

After the last of the fires had been extinguished, the landscape was apocalyptic. Sara is shown in *Fires of Kuwait* wandering among the charred ruins, smoke from the smoldering ashes shrouding her as if it were mist rising from the ground. "It's not what you lose on the surface," she says, her voice tinged with sorrow, "it's what you lose underground now, and that's a tremendous amount. Each and every well is like a patient that you know very well, like a human. We deal with them when they have a problem and we try to solve that problem" (1992).

Reflecting upon the environmental devastation caused by the oil well fires and the resultant pollution to the land, air, and water, she says, "To me, working in the field is a very peaceful place, very calm environment that I really enjoy with all the flowers, all the grass, all the green, birds, animals, insects. After all these blow-outs, it's no longer the same, and I think it will never be the same" (*Tides of War,* 1992). Twenty-five years later, Sara visits the fields and sees how the Earth has managed to repair itself almost to its natural state. "This tells me how little we know about our ecosystem," she says.

In 1993, the United Nations presented her with its Global 500 Award in recognition of her heroic efforts to protect and preserve the environment. She would later receive the American Institute of Mining, Metallurgical, and Petroleum Engineers (AIME) Charles F. Rand Memorial Gold Medal in 2013 for distinguished achievement not only for her accomplishments in the field of petroleum engineering, but also for her "wisdom, bravery, and leadership that made her a hero during the first Gulf War."[18] She was the first and only woman, to date, to be given the honor in the 75 years that it has been awarded.

Less well-known, perhaps, than her unique role as the only female firefighter is the extraordinary courage she – and others - displayed in

---

[18] Source: AIME

smuggling the records out of Kuwait Oil Company, while Iraqi guards patrolled outside.

"KOC's essential information and data base was all contained in microfilms," she explains. "It was very dangerous, but we thought that it was worth it. We took the risk because we knew what was involved." Sara and the other individuals involved hid the records for safekeeping – in Sara's case, in her bedroom – and returned them once Kuwait had been liberated.

Seven months after her work on the Kuwait Wild Well Killers team began, it was time to start rebuilding Kuwait's petroleum industry. Sara was promoted to head of the petroleum department and spent the next two years working on the rehabilitation of the wells that had been damaged.

As her career at KOC progressed, she moved into various aspects of management, such as planning and resource management. After 18 years, looking for a new challenge, she then went to work for Kuwait Foreign Petroleum Exploration Co. (KUFPEC) in 1999 as manager of new business development. Six years later, feeling like she had climbed the corporate ladder as far as she could, she established Kuwait Energy Company, a multi-national, independent oil and gas exploration and production company. Although the company was founded in Kuwait, its E&P activities take place in other Middle Eastern countries, currently Egypt, Iraq, Oman, and Yemen.

Looking back at the last 10 years, having been at the helm of her own company, Sara says she has enjoyed every aspect of it from making investments, going into new markets, dealing with financial institutions to employing personnel, building a team, and being committed to social work – "having this *drive* to actually better the lives of people and the environment" – all of which has been very satisfying to her.

Despite the personal fulfillment she has derived from being an entrepreneur, neither she nor her company is immune to the vagaries of the international commodities markets. During the current extended downtown, which has seen the price of oil drop by ~60%,[19] her company is facing challenges just like any other. With more than three decades in the industry, Sara is well aware of its cyclic nature, but says, "Unfortunately, we as governments, as industry, have not learned from the past, and this pattern will continue," unless different strategies are employed.

---

[19] This figure reflects the drop in the price of oil at the time this interview was conducted, not at the time of publication.

She believes there needs to be as much collaboration as possible between governments and industry to remove the extreme volatility in pricing, which has created a situation that "is really not beneficial, not to the governments and not to the consumers." In her opinion, it is even worse for the industry because it results in a loss of human capital and, when the market improves, there will be a need to re-hire amidst a possible shortage of experienced workers.

Downturns also lead to investments being cut, which means "the oil price will shoot up and everybody will invest, then it will have to come down again. There *must* be a better way of managing this oil price," she says emphatically. "Oil is a strategic commodity and everybody uses it. It's really wrong for the world to manage it the way it is doing now. It's hurting industry, governments, countries, and this is not good for anybody."

When she is not dealing with fluctuations in the world oil market, Sara finds herself confronting another challenge, the same one she faced over 30 years ago – women in the field. Despite being a vital experience for them to gain the critical skills that will enable them to advance in their careers and later into leadership, there is still familial and cultural resistance to the idea, even from within her own company.

Sara remains determined to bring Arab women into the industry. She has partnered with SheWorks, an International Finance Corp. (IFC) initiative that encourages a "gender-smart" workplace culture that encompasses mentoring, sponsoring, and leadership development. While she sees signs of improvement, she wants each country to employ at least 30% women in all disciplines. "That's the kind of diversity we're trying to bring to each location.

"We still struggle, even at Kuwait Energy, for men to allow women to apply for these jobs and to encourage them to take these positions. It is only when I push and force them that they accept these kinds of measures allowing women to take responsibility in fields."

Currently, she employs two female petroleum engineers who are working in the field in Iraq and three or four in Egypt, but she has not been able to get any women engineers into the field in Yemen, where KEC has operations. She acknowledges they have women in administrative positions, "but I'm talking about leadership positions and technical positions. That's where I encourage women to go and learn and participate and be in the field. I think that's an opportunity they have to have."

Sometimes she is faced with the argument that "quality women" are not available to fill the positions. While Sara agrees that quality must come first – "I cannot jeopardize operations just because I want to hire women" – she believes if there are two equally-qualified candidates, a woman and a man, the woman should be given the opportunity. With women making up about 70% of enrollment in Kuwaiti universities – and disciplines like engineering attracting large numbers – employers would certainly seem to have a highly-educated talent pool of female candidates from which to select.

Once they do join the industry, Sara says women prefer to cooperate as opposed to competing, something she believes is attributable to both a woman's mindset, as well as corporate culture. "It is the nature of women to build bridges rather than trying to compete all the time."

She sees cooperation as a strength, as well as one of the keys to success and even survival. "For a company like ours to grow in the Middle East, we have to cooperate with everyone – the majors, national companies, governments. We cannot compete with giants – the BPs, the Shells, the Exxons of the world – they are too big and too well-established. It's suicidal to compete with them. Through cooperation and goodwill, we have an opportunity to get projects."

As difficult as that may be, particularly as a woman in a man's world, Sara firmly believes in "the power of women." For any woman to work in the petroleum industry, she says a woman normally has to put forth twice the effort of a man, in order to prove herself. At the same time, she has observed a woman will go the extra mile to ensure things are done on time and on budget. "In that sense, women are far more powerful in execution than men. I have seen this in many parts of the world, but more so in this part of the world."

She asserts that women, by nature, are far more patient than men, and have more emotional intelligence, including empathy, and are generally more passionate about their work.

"Now, if you combine all these elements, then you have a very powerful mix – for anyone." But, she feels some of the traits are more innate to women than to men and "that's why I say when women are given the positions, they are far more powerful in what they do than men."

Sara implores women not to limit themselves and reiterates that when they believe in themselves, when they have passion, they can be very powerful. At the same time, she notes that women are risk-averse and prefer to take the "clear road, clear way." And on this point, she is very clear: "In this business, I'm afraid they have to take risks."

Not any risk she stresses, but calculated risks. By avoiding risks she warns women limit their opportunities to progress. Risks are linked to rewards and, if women want those rewards, they have to be willing to take chances.

"The tendency of women is to be risk-averse. This is my advice: They have to be aware of that, they have to overcome it, and they have to take some risks," says the woman who was willing to walk through fire to get where she is now.

♀ ♀ ♀

In December 2017, Sara Akbar resigned as CEO of Kuwait Energy. Effective January 1, 2018, she became a non-executive director of Petrofac. She is a Member of Higher Planning Council – State of Kuwait and Trustee of The Silk City and Islands development – State of Kuwait. In September 2018, she was inducted into the Offshore Energy Center Champions Hall of Fame in Houston, Texas.

5	**Jerry Tardivo Alcoser** **Petroleum Engineer**

"I think the biggest thing is, don't rule yourself out."

What's the worst thing that can happen? That question has become the litmus test through which drilling and completions engineer Jerry Tardivo Alcoser filters her decisions in life, including – maybe especially – those that pertain to her career.

Jerry grew up in the rural town of Apollo, not far from Pittsburgh, Pennsylvania. This southwestern area of Pennsylvania was a Mecca of steel mills and coal fields. Her father, Dave, became a coal miner shortly out of high school and when lay-offs happened, as they often did, he utilized his skills in masonry. As a youngster, a strong work ethic was the only example Jerry was shown. This "make it work" attitude was something that influenced her more than she would know. Her mother, Megan, stayed at home and would "fire up the coal furnace" to keep the small farmhouse warm when her dad worked midnight shifts. Jerry has fond memories of playing on the big heaps of coal with her sister, Andrea, in their basement.

She also remembers tip-toeing around the house in the afternoon when her dad was trying to rest before leaving for a long night of work under the ground. Family was always the most important thing and "we would talk about our day's events with one another every evening at dinner." Jerry's mom would later return to school, earning a medical secretary degree, and re-enter the workforce to help contribute financially when her girls were old enough to enroll in school.

"We would just make it work as a family," Jerry says simply. "Growing up, we didn't have a lot of money, but we had everything we needed. We started in a four-room farmhouse that my dad had purchased on some acreage, and we remodeled the house together. My sister and I would help him. I've done everything from laying brick, and striking and raking the mortar joints, to putting in hardwood floors."

She learned two very important lessons from observing the cyclical nature of the extractives industry – saving money and having a back-up plan. "My dad is absolutely brilliant when it comes to math and science, and problem-solving. I could always tell how passionate he was about being able to create something, whether it was pouring a foundation or building an entire house. When he wasn't working in the mines, he had several other skill sets that would bring in money."

An unforeseen bonus? "Doing all that work with my dad was great for what I'm doing now, having gone into engineering, having a mechanical-type mindset. I'll be honest, I don't know if my dad thought it was a possibility for me. I'll never forget, when I was looking at colleges, I told him, 'I think I'm going to go into petroleum engineering.' And he said, 'Jay, that's a man's field. Why would you want to do that?'"

Jerry believes he was equating the petroleum industry to the coal mines, where he witnessed the difficulties faced by a female colleague, whom he would stand up for when other miners made disparaging remarks. "Granted, this was 25 or 35 years ago. But she had to make a living for her family and my dad was very supportive of that. I knew he was thinking, 'I don't know if I want my little girl in that. Are you sure that's what you want to do?'"

It was. In 1998, she enrolled at Marietta College, a small prestigious private college in Marietta, Ohio, known for its petroleum engineering program. It is also located in Washington County, an area that experienced oil booms in the late 1800s and early 1900s, and, more recently, was found to lie within the Marcellus and Utica shale trends.

During college, she worked for a small oil and gas company, reporting directly to the president, doing reserve and decline-curve analysis. She implemented a new software system to forecast the reserves that the company still uses today. The president of the company became a mentor, offering beneficial insight as Jerry was making future career decisions. She still refers to some of his wisdom and knowledge today.

On her summer breaks from college, she gained valuable experience that would help shape her career when she secured internships with different oil and gas companies.

Her first internship after her freshman year was with Mobil Oil in Liberal, Kansas, located in Seward County, the site of the vast Hugoton natural gas field discovered in the early 1920s. It was the first time she had ever been away from home other than on family vacations. It also was the first time she had ever eaten Mexican food and heard people speaking Spanish. Jerry calls the experience "a great diversity moment."

The next summer, she worked as a reservoir engineer for Marathon Oil in Houston, Texas. In this position she performed reservoir simulation analysis for Gulf Coast offshore operations. She simulated various water-floods and ran project economics to determine the best path forward for the project. She also didn't have a car, so she chose to live close to the Marathon building and recalls sweating in the Houston heat walking to and from work.

Her last summer before graduation was spent offshore in the Gulf of Mexico (GoM), as an intern with Chevron, the company that would become her future employer.

It would be her first experience with helicopter underwater egress training (HUET), which is required before going offshore. "I was amazed at how disoriented I got when flipped upside-down under water. Not to mention water rushing up your nose. The six of us got out by just trying to stay calm. The training is absolutely critical."

Jerry's training took place at a time before the industry really recognized the need for women's clothing sizes and the insulated survival suits were one-size-fits-all. "It's me and forty guys in this water survival safety class. I'm trying to get into this cold-water suit and I could barely maneuver because it's so big. Then I had to climb out of the pool using a ladder and my feet were at the knees. I looked at the instructor and asked, 'Do these come in petites?'"

Another component of HUET involves swimming while dragging an "injured" colleague to safety. "I figured I got the biggest guy ever. I was like, 'I don't think this is fair. He probably weighs four times the size of me.' But we did it!"

Yet another test is jumping off the high dive. "There was this 6'4" guy that was scared to death to jump. So, I was thinking, 'Okay, he is panicking, and I can jump off the high dive, that's not a problem.'" Her observation? "You see offshore survival training is very gender-neutral."

It also was during this time that Jerry met one of her most influential mentors in the industry, Jaime Crosby, who would remain a lasting friend. He was a drilling superintendent and her first mentor when she was a summer intern. He continued being a mentor and would become her first supervisor at Chevron.

After graduation, she once again sought his advice when she was debating between two full-time job opportunities – one with Chevron, the other an international role with a different company. After talking for an hour and a half, he told her, "Ultimately, it's what you want to do," which helped Jerry come to the conclusion, "If I start in an office, I'm going to end in an office; I'm never going to get field exposure experience. Might as well give it a shot. *What's the worst thing that can happen?* I don't ever want to look back and say, 'Could I have done that?'"

Another piece of advice from her mentor: "Before you manage the store, you've got to learn how to stock the shelves." Jerry says she uses the quote even to this day with her mentees. "It's so true. In any type of industry, if you can understand what's going on in the field where the work's being done, it really helps your perspective as you move into leadership and management roles."

Ultimately, she chose the position with Chevron and has remained with the company throughout her career, which now spans nearly 17 years in the industry. In that relatively short period of time, she has worked her way up to her current position as business development and planning manager for the Deepwater Exploration & Projects (DWEP) Business Unit of Chevron North America Exploration & Production Co. She is often asked about her rapid climb up the corporate ladder, to which she responds, "Opportunities were presented to me and I said, '*What's the worst thing that can happen?*' I could fall flat on my face and then I'll go back and do something else, but you're never going to know unless you try."

Which is exactly how she found herself working offshore as the company man – "company person, actually," she corrects herself – in the Gulf of Mexico shortly after hiring on with Chevron in 2002. Because there were relatively few female company persons at that time, Jerry says at first the rig crew struggled with her being on location. In time, through working long hours and problem-solving together, something changed. She noticed an interesting dynamic had developed. "We were a *family* because we spent a lot of time out there together. Granted, for the first two or three months, the guys don't really talk to you because they're scared to death of saying the wrong thing and they don't really know you, right?" Over the years, however, Jerry has built long-lasting friendships with the people she has worked with on the rigs.

She relates the story of the first time she had "broken out, working days, leading the rig by myself under the mentorship of Kyle Harrelson. He gave me the keys and said, 'Call me if you need me.' Kyle has provided me with insight into everything from drilling wells to production forecasting. He has been a rock for me throughout my career; he is truly family." Kyle was the first to meet her now-husband, Luis, in Angola, Africa, where all three were working, and he also attended their wedding.

Jerry remembers when Hurricane Rita was bearing down on Houston in 2005. Kyle and his family, including his wife, mother, and four kids under the age of five, went to stay with her in Lafayette, Louisiana. The drive, which normally takes four hours, took them 27 hours in two cars.

"I was single at the time and my house definitely was not child-proof. I remember looking at Kyle and his wife, Heidi, when they showed up and telling them, 'Child-proof as you need to.' I went to the grocery store and, after waiting about three hours in the check-out line, I came home to a baby-proof house and a big family to share dinner with. We were all safe in my 1,400-square foot house throughout Rita."

After three years working a 14-days-on/14-days-off rotation schedule, Jerry came onshore and for the next two years worked as a drilling engineer for the Gulf of Mexico. During that time, an opportunity arose and she was offered a resident job in Kuwait. This is when she learned something she now shares with colleagues and mentees: "Don't be afraid to say no." Her late grandmother was an international stockbroker with Merrill Lynch and had lived in Bahrain, so Jerry was familiar with the Middle East, but it was not a posting she wanted to take as an expatriate resident single status.

She did accept the next international offer, working a 28/28 rotation schedule offshore in West Africa, as Chevron's first woman in the drilling and completions office in Cabinda, Angola. It would prove to be a good move, not just for her career, but for her personal life, as well.

It was in Angola that she met her husband, Luis Alcoser, also an engineer for Chevron, who was working completions while she was working drilling. "He kept asking me out and I said, 'No way.'" Apparently, engineers can be persuasive as well as analytical. She finally relented and says, "We went on our first date and haven't been apart since." They began rotating together and were married within a year-and-a-half.

Jerry admits she hasn't always had her current level of confidence and that earlier in her career she was more of a worrier. It was during a discussion about job moves that her husband first asked her, *'What's the worst thing that can happen?'* She says, "He is very supportive. He has truly been my sounding board and partner in all of my adventures."

Returning stateside in 2009 after two years in Angola, the couple settled in Houston, Texas, where Jerry began working for Chevron's Vice President of Drilling & Completions, Dave Payne, whom she describes as a great leader and someone she considers a mentor. They were having a career development conversation – "Chevron does a phenomenal job with career development plans and discussions" – and he asked her what she wanted to do.

Jerry told him she would be very interested in an engineering manager job. The Mid-Continent Business Unit, Chevron's US land group, had an opening and Dave asked her, "Why not that one? Why not now?"

Jerry remembers ticking off a list of reasons in her head why she shouldn't take the job – "I had never worked land, I had never supervised people" – and then asked herself the all-important question, *"What's the worst thing that can happen?* If I completely fall flat on my face, then I go back to an engineering role or I do something else, right? But you're never going to know unless you try. I said, 'I was thinking maybe five years from now, but let me go talk to the drilling manager.'"

The other person she consulted was her husband. "He was extremely supportive. He said, 'You can do this! Why would you even think of not taking it?' It's great having that kind of reinforcement at home," she says. "Truly, I wouldn't be where I am today without him." In accepting the position, she became the first woman to hold a drilling and completions engineering manager role at Chevron.

As her career has progressed, Jerry has noticed a significant change in the industry. Companies like Chevron have seen and appreciated the importance of inclusion and diversity, and today there are a number of female asset development managers, petro-techs on the platforms, and women holding a wide variety of positions both offshore and on.

Less than six months after Jerry became a drilling and completions engineering manager, Chevron made an external hire and brought in Kim McHugh, who also would break barriers as its first female drilling manager, and become Jerry's supervisor, mentor, and friend. "So here you had both of us working in the business unit," Jerry says, calling Kim by far one of the best leaders she has ever worked for.

"Miss Bonnie has been another great mentor as well," Jerry says, referring to Bonnie Maillet, CEO of Boysenblue Celtec, in Lafayette, Louisiana, whom she met while working for Chevron in the Gulf of Mexico. Their companies did business together and the two began building a relationship, serving as sounding boards and offering each other support. "She just tells it like it is and that's one thing I love about her." They have stayed in touch over the years and Jerry considers Bonnie family.

Because she has been fortunate to have such good mentors – both male and female – Jerry is committed to sharing what she has learned in nearly two decades in the industry with the up-and-coming generation. As the mom of a three-year old daughter, Eva, she says, "Something I'm very passionate about is getting in high schools and educating women – and men, as well – that daughters can go out and do those certain things because, as I said, I don't know if my dad thought it was a possibility for me."

Jerry does mentoring through the engineering leadership program at her *alma mater*, Marietta College, and has established mentoring and networking groups at Chevron. "I hold a quarterly mentoring session within Chevron – men and women, various disciplines and age ranges – so that everything is on the table. People bring up all sorts of things, their experiences and challenges. I am so passionate about mentoring, making it easier for people and inspiring them to go do things they want to do.

"I think I get more out of mentoring than the mentees do, honestly. They speak very frankly, which I like, because we set up the discussion to be like Vegas ('what happens in Vegas stays in Vegas'). It gives me insight into what's going on in the business and what they're dealing with." She finds that mentoring provides an avenue to break down barriers between upper management and its organizations.

Now at 36, Jerry appreciates the perspective of the younger generation, even though it can be somewhat disconcerting to realize she is no longer the youngest person in the room.

"People just coming out of school think differently than I do. Their perspective is different. It's a completely different generation. I stepped into the elevator the other day; there were these guys that didn't even look old enough to drive but they were employees. I said, 'You know what? It's officially happened. I'm no longer the young one. I'm now the one who says, Hey, remember when? Well, I can tell you, when I was offshore...' I said, 'Oh, no, it's happened.'

"I truly think you're seeing this shift in the industry because more women are going into STEM. The young millennial generation especially is dual-career focused and I'm seeing some tremendous talent. Very passionate and very intelligent women coming up, whether it's through Chevron or industry in general, so I'm very excited about that going forward.

"I think the biggest thing is, don't rule yourself out. I get the question all the time, 'How did you handle being offshore? That's so difficult.' Granted, it may not be for everybody, but you've got to try. Sometimes, I think the age gap can be more of a challenge than being a woman."

Despite her personal philosophy and healthy self-confidence, Jerry admits to having experienced some doubts when she first went offshore. She questioned how she was supposed to lead, especially given that she was just out of school. She wondered how her academic knowledge of the industry would translate into real-world experience.

"Sometimes, people in general – not just women – will rule themselves out. They say, 'I can't' and they don't even try. You'll be surprised what you can accomplish when you just jump in with both feet. People love to tell you why you can't do things. I say, 'Tell me how you can. We'll make it possible.' You've got to ask yourself, *What's the worst thing that can happen?"*

♀ ♀ ♀

As of this writing, Jerry is now general manager of operations at Chevron's Bakersfield, California, location. In October 2018, she and husband, Luis, welcomed their second daughter, Gianna.

	Ann Cairns
6	**President, International Markets, MasterCard Worldwide**

"One of the things about being a woman in
a man's world is that you're very noticeable."

No one who knew Ann Cairns growing up in Newcastle, in the northeast of England, would be surprised that she is currently president of international markets for MasterCard Worldwide. Technology, credit cards, banking, numbers, maths. It makes perfect sense that the young girl who grew up in a family of mathematicians – father, mother, older brother, and younger sister – would be involved in the finance industry. It was practically destiny. But could they ever have dreamed that she would get there by way of the oil and gas industry? That takes a slightly greater stretch of the imagination.

Both Ann's parents left school at the age of 14, and, despite not being formally educated beyond that age, she describes them as "naturally" good in maths. Her brother John, who is a technologist and mathematician, and her sister Kath, who has a computing and pharmacology degree, have a fantastic aptitude for maths as well. "I think some families tend to have similar skills," Ann says, offering an explanation as to how all five of them could be so mathematically minded. "Also, my brother and sister are both musicians, and often musicians have mathematical skills or vice versa. Unfortunately, I wasn't much of a musician. I can sing a little bit, entertain at the Christmas party," she says, laughing.

Her brother, who is older by four years, has "a very scientific mind," and, as kids trying to fall asleep at bedtime, would instruct Ann, who was about seven, to "calculate the number of seconds in a century and don't miss the Leap Years because I'll know." She breaks into peals of laughter at the memory.

Looking back, she recalls she never particularly liked dolls or "girlie" things of that nature growing up. For Christmas, she would get post office sets as gifts and would then set up her own post office, place postal orders, and count the play money. "Sounds weird, doesn't it? But, those are the kinds of things I used to love to do – of course, card games such as Canasta and board games like Monopoly and Risk were appealing."

At the time, Newcastle was a large industrial town with an economy based on ship building and the extractives industry – in this case, coal mines – although that would change later as the mines closed and more service industries moved in. It is also famous for its football team – "not doing well lately," Ann points out.

Her family lived near the city center until she was about four, and then they moved to a mining village where almost all the men worked in the coal mines. "It's probably a bit like living in Pennsylvania," she muses.

Her father Tom however, was a shoemaker. He started out making clogs and footwear for the miners. Then, as mass-produced shoes became more commonplace, he went into repairing shoes. Eventually, he began focusing on custom-made shoes for people who had difficulty walking or specific problems with their feet – "the medical side of it. It was quite interesting."

Although she grew up around male-dominated industries, Ann says, "My father was an unusual northeastern man in that he said to me, and also to my sister, 'Look, you can be anything you want to be when you grow up. Never say, I can't do something because I'm a girl.' He was very encouraging from that point of view.

"Remember, at that time, it was more normal that the father was the person in the household who worked, who brought in the money, so they had very dominant positions. The northeast was a very male-dominated environment when I was growing up, so for your father to say that to you was important."

Armed with that kind of paternal support and validation, Ann went on to earn a first-class bachelor's in maths from Sheffield University and a master's in statistics from Newcastle University.

"I actually gravitated toward pure maths and statistics. I like them both equally. People say to me, 'What's the point of pure maths?' And I say, 'Well, it's interesting. I think it describes the universe, doesn't it?' I particularly like things like probability theory, and that dovetails with statistics.

"On the one hand, statistics is a very practical subject, and you can write essays in it – not that I did a lot of essays at university (laughing) – but you're analyzing a lot of real data and you're trying to create empirical models and so on. On the other hand, pure maths is much more artistic, it's describing often abstract concepts in a way that can be understood, so that everyone would understand the rules by which the universe operates, there's an infinite number of infinities – and their sizes would surprise you! There's an infinity between zero and one, and then there's another infinity between 0 and 2, and those two infinities are actually the same size Well, actually, the infinity between 0 and a million is the same size too..." she stops and laughs. "No, I better not go down this track. It makes you really think, doesn't it?"

After graduation, Ann was hired as the first female engineer to work alongside physicists, chemists, and mechanical engineers at a research station in the gas industry. The experiments focused on the pipes and determining how to detect what was under the ground without actually digging it up. Pattern recognition capabilities were used to bounce signals into the ground to locate the gas mains, electricity cables, and water mains, so the gas company would know where it was safe to dig up the road.

"If you've ever seen any of the big pipes that bring in the gas from the North Sea, they're green-colored. They've got this epoxy-powder coating on them and the chemists decide what temperature they should heat that coating up to, so that it sets in a way that's not too brittle, but not too soft, and creates a very strong protective environment around the pipe. I would design the experiments working with the chemists that would test those kinds of things.

"Also, in order to determine how strong the steel is in the pipe, we would do things like test pipes until they exploded. You could test at what pressure the pipe would start to unzip itself and then you knew within certain tolerances what the network could withstand across Britain and so forth. Coloring them green, blowing them up... I knew where the pipes were in the ground. It was absolutely fascinating. I loved it."

When she wasn't setting off explosions, Ann was honing her pyro-technical skills – all in the name of research, of course. "Another experiment we did was set a big container of liquid natural gas (LNG) alight to see how the flame moved, what heat intensity it was, all of these things. You have to do that in a big open space, up in the hillside somewhere, away from populations. Some of the experiments took place in a town called Killingworth, which is up in the northeast of England, although we would do experiments all over the country."

After five years, Ann had had enough excitement on dry land, and turned her attention toward the offshore industry, which she viewed in the mid-80s as "cutting edge."

"Really, when you think about it, that was where the gas was coming from, if you like, and there was a lot of investment in that side of the industry. You were going right to the source of the fuel. And, of course, you were working in very difficult conditions, so in the sense it was testing science, trying to build the rigs in the right way, trying to maintain them properly, trying to find the oil and gas in the first place, so that was the exploration part."

Even though she had no specific job in mind, she looked up Bruce Goodwin, then-head of offshore engineering in British Gas – "South African, very interesting chap, actually" – in the phone book and boldly called him. "I rang him up and said, 'I'm a research scientist up in Newcastle, and I'd love to work offshore.'" He was intrigued by her work and invited her to travel to London to meet him for lunch to discuss the possibility of a role offshore.

She took the train to London and during their meeting Goodwin told her British Gas had a "Scottish chap" working as a consultant for them, running the offshore engineering planning area. He suggested Ann study under him as she worked her way up in the planning department. She accepted the position and a mere three months later, the consultant left, leaving Ann to take over. "It was a bit of a short apprenticeship, to be truthful about it." Apparently, it also made Ann the first female engineer certified to work offshore in the North Sea.

Prior to going offshore, she spent the next few months qualifying to pass the safety training. The dry suit she was issued was too big for her small frame and, one morning on the way to work, she accidentally left it on the train. She went into Lost Property in Waterloo Station and said, "I've left my bag on the train from Bexleyheath."

"What was in it, madam?"

"An orange rubber suit and a tin of powdered milk."

"And they went, 'Really?'" She bursts out laughing at the memory. It was never returned.

The actual training itself was nothing to laugh about. "You're in a simulated helicopter and they say to you, don't actually undo your seatbelt when the thing is falling into the sea because otherwise you will churn like washing in a washing machine, and then you won't know which way is up. It takes a great presence of mind to sit with your seatbelt on when you're drowning," she says in an understatement.

"First of all, you hit the water with a bit of a bang and it knocks all of the air out of you, and then it's dark. You've got a desperate desire to let go of the seat buckle and swim somewhere. It's an instinct that's almost overpowering, but you've got to *stop* yourself doing that.

"Actually, you have to do [the simulated helicopter ditch] four times. There were four of us – three guys and me – and we rotated, so sometimes you were the first out and sometimes you were the last out, which is horrible." All the while, water was gushing down the back of her poorly-fitting dry suit.

The fact that the trainer was an "ex-Royal Marine kind of guy" didn't help matters. Ann was the lone woman in a class with 47 men and the instructor kept insisting she go first.

"I was fairly determined that I wanted to do the job, so it didn't really matter that the guys were watching. I just got a bit miffed at him because I thought he was singling me out. It was only on day three that I got round to saying to him, 'Why are you making me do everything first?' He said, 'So that the guys following behind you have just seen a small woman do this. It will make a big difference to them. It will be easier.'"

She laughs, "He was using psychology. At first, I thought he was just being unfair. Actually, when he told me that, I was perfectly fine with it. The thing I was worried about the most was actually driving the lifeboat. You know, those 45-men lifeboats that are orange and they kind of dive into the water off the platform and then they bounce and you drive off in it," she says, making it sound like something from a James Bond movie. "It was a classic thing – I thought they were going to go, 'Agh, women drivers,' because I've got to say, I am not the best driver on the planet. I do sail, though, so it's actually funny that after all this time I still love being on the sea."

Once offshore, "I was involved in the production part, which was basically, once you build the platforms, how do you keep them safe, how do you keep them comfortable, how do you stop them having all sorts of rusting problems, and problems with the whole marine biology life, all of these things, and how do you make sure that all of the equipment is in tip-top condition?

"We had three big platforms – two in the southern North Sea and one in the Irish Sea – that we were maintaining and all of the planning programs, the way that you would plan the work for the guys to do on the platforms – the dismantling of the compressors, the replacement of the topsides. Well, if you think that I've got a mathematical back-ground... these programs were optimizing the way you did all the planning schedules and so on. Then I would travel out to the different platforms and see how all that was going, and whether what we were planning was actually feasible, or were there different ways we could be thinking of this. So, I was learning on the job how everything worked on the platform."

Unlike the majority of offshore jobs, it wasn't necessary for Ann to work a rotation schedule. She would go offshore and stay on the plat-form anywhere from three to five days, which enabled her to join the senior management meetings. While at first she liked the idea of not working a rota, she quickly discovered it had a downside. "Actually, I think it was a worse situation because I would have to go back and start working in my office in London, so I didn't have the nice breaks I would have had, if I were permanently on the platform."

She found the men on the platform to be very enterprising. "Obvi-ously, they had to occupy themselves when they weren't working. Some of them were running the London Marathon, so they'd be training on the helipad. Others were doing Open University degrees. I remember one guy, who was running one of our platforms, was doing an English and history degree with the Open University, so he had lots of books and essays to write and so forth. Very interesting."

While she concedes that not everyone was studying, "so there would be other types of entertainment such as 'cinema,'" Ann has been quoted as saying, after moving into banking in the late '80s, she found it to be more sexist than the petroleum industry (Harlow, McKenzie, 2015). However, she clarifies that by saying, "In the engineering world, people are very collaborative. Everybody's got their own skill base and you tend to work in teams. You're valued for what you do rather than what gender you are. People appreciate what you bring to the party and it's not questioned in the sense of gender; it's just 'Are you good at your job?' If you are and you're collaborative and you help build the solution, then you're just accepted. Whereas when you go into, say, the investment banking world, people work quite a lot by themselves."

Despite the emphasis on teamwork in engineering, as one of the first female offshore engineers in the UK, she was often the lone female team member.

"There weren't other women," she says matter-of factly. Then she adds something rather startling, in light of today's emphasis on mentoring. "I don't think that you necessarily need female mentors.

"When I was a research engineer, Margaret Thatcher was the first female prime minister of the UK.[20] She had studied chemistry at Oxford; she was a scientist. She came up to visit the engineering research station where I worked. I would have been in my very early 20s – 22 or 23 – and I wasn't a follower of the Conservatives at that time."

Ann admits that her opinion of the Prime Minister had been influenced by the media, particularly newspapers, and that she came across as a very powerful and opinionated person. However, actually meeting Margaret Thatcher was a big surprise. Ann found her to be extremely charismatic, giving each person her full attention while she was speaking to them.

"I was a mathematician and statistician at that time and she would talk to you about your work and, because she had a scientific background, she was able to engage you. And so while she wasn't a mentor for me, after I met her I saw her as quite a role model because she was somebody who had grown up in Britain, had gone to university, had been a scientist, and had risen to run the country. You couldn't help but be impressed by her whether you agreed with her politics or you didn't."

One of the benefits of being surrounded by male colleagues meant there was no shortage of male mentors and Ann says even today she has many mentors who are men and have been either direct bosses or senior members in the organization.

"One of the things about being a woman in a man's world is that you're very noticeable. And I got quite good exposure. I was always very eager to ask for people's help and advice. So I built quite a strong network across the whole business of men that helped me in my career actually."

Although it was some years after her engineering career, in 1993 Ann became pregnant with her daughter, Sophie, while working for

---

[20] After Margaret Thatcher's three terms in office (1979 – 90), there would not be another female prime minister of the UK until Theresa May's election in 2016.

Citibank[21] and her boss Bill Grant actually promoted her while she was on maternity leave.

"That just goes to show you how supportive men can be and how much faith they can put in you," she says.

She also cites Bruce Goodwin, "the South African chap," who gave her the first job offshore, which culminated in her role as one of the first female offshore managers in the UK. She led 50 or so engineers involved in the monitoring and maintenance of platforms in the North Sea and the Irish Sea.

"You have to dismantle the compressors; refurbish the topside; [carry out] diving programs to make sure that the undersides of the platform are safe and you don't have marine organisms attached that are fatiguing the structure. There are continuous maintenance programs throughout the year. And there are the logistics – you've got to have support vessels, and spares flown onto the platforms, and so on – which involve computer-based planning. The department that I ran was responsible for all of that and I would fly offshore to the platforms to make sure that the plans were being followed."

Up to this point, British Gas Corp. had been the publicly-owned monopoly provider of gas in Britain; however, in 1986, under Margaret Thatcher's Conservative government, the gas industry became privatized and British Gas plc. was formed.[22] Prior to that, Ann says the company had been run "predominantly by engineers" and, in fact, the CEO, Sir Denis Rooke, had a mechanical and chemical engineering degree from University College London.

"During privatization, it was becoming clear to me that the engineers weren't the people who were taking over the key functions in the company; actually it was people with financial backgrounds or at least that's what it looked like to me. Remember, I was in my early 30s so I thought, my goodness – and this is a classic thought, isn't it? – if I want to run this company, I should have some real financial training and maybe I should step out now and go to London Business School, do an MBA."

Fortuitously, Ann spied a Citibank advert in the *Sunday Times* newspaper, recruiting "mid-career hires," people, she says, in their early 30s, from various industries who had management experience. It so happened that the CEO of Citibank at that time, John S. Reed, had an engineering

---

[21] Ann's career with Citibank spanned 15 years (1987 – 2002).
[22] Source: British Gas

background with degrees from Massachusetts Institute of Technology (MIT).

"They were particularly looking for people with the kind of background that I had," she says, deciding it was too good an opportunity to miss, despite not knowing anyone at Citibank or having connections in the banking world. "There wasn't anybody I could ring up and ask to recommend me." Out of thousands of applicants, 800 were shortlisted, and of those, Ann was one of 25 chosen.

This led to the aforementioned 15-year career with Citigroup, which was followed by a six-year stint, beginning in 2002, with the Dutch bank ABM AMRO.

Ann says the bank was interested in sustainability, ethical investment, carbon trading, and had looked at investments in clean energy companies.

"In their buildings in Amsterdam, it was interesting to see lights switching themselves off, blinds going up and down at certain times, the heating modulating so that it didn't burn energy unnecessarily. The Dutch were way ahead of everybody else."

She then became managing director of Alvarez & Marsal in London during which time she led Lehman Holdings' business across Europe through its first two years of Chapter 11. Other clients included Icelandic banks and the Irish government.

After more than three years with Alvarez & Marsal, she embarked upon her career with MasterCard Worldwide in 2011. Although the company moves billions of dollars across the globe in mere seconds, Ann says MasterCard is not a financial institution; it is a technology company. Its focus, investment, and money are spent on technology, she says, citing its futuristic biometric authentication.

"We're very focused on financial inclusion. Millions of women are born all over the world and, particularly in parts of Africa, their births aren't registered. They don't have passports, driver's licenses, bank accounts. It's hard for them to be noted anywhere as people in their own right.

"Once we start to work with governments rolling out capabilities to the general population which allow them to biometrically authenticate themselves with fingerprints, voice-recognition, heart rate, iris-scanning, and so on, suddenly, even in these kinds of geography, they have a digital means to say, 'This is me.'

"This is engineering again because biometric authentication is kept in a chip on a card or increasingly in the phone. And by being able to rollout that capability, you're giving hundreds of millions of people – a high percentage of them women – the ability to have a digital identity in the world. It's hugely, hugely impactful. And it's absolutely all to do with engineering."

Although Ann says MasterCard employs quite a few female technologists and engineers, its program Girls4Tech works with girls as young as 13 and 14 to expose them to the industry.

"As I'm talking to you, I'm actually looking at a picture of myself with a group of girls that we've brought into our office here in London. But we also do it in India and America and the Middle East. It's so important to build a pipeline of qualified women for the future. If you have diversity, not just from gender point of view but from all sorts of different backgrounds, you create groups of people who have different skills in order to actually build something stronger. I always laugh and say, 'Who would want to walk across a bridge built by bankers? Least of all bankers!'" It is a joke she is eminently qualified to make.

Ann sees great potential for the future of the energy industry – and the opportunities that presents for women – as it expands to include a number of alternatives to fossil fuels. "If you're a very value-based person and you're interested in sustainable energy, clean energy, and so on, there's a huge amount of investment done by the energy industry in this on a global basis, which could be very attractive."

As someone who was at the forefront of women gaining entrance into the oil and gas industry and then utilized those skills to transition to a completely different sector, Ann advises young women to acquire a wide range of skills, particularly ones that can't be easily automated, and not be afraid to step outside their comfort zone, and even make career changes without worrying about climbing the corporate ladder. She believes the right way to think of the ladder is as a long extended journey of one's life and to be open to as many experiences as possible.

"We should be making people feel very confident and comfortable about what they can achieve in life regardless of what their background is. And I actually think that it's not just the work environment that generates that interest. If I think about myself as a kid – you know a first man walked on the moon when I was 11 years old – I remember going outside and looking up at the sky and thinking, 'Oh, I wonder if I should become an astronaut? This is very exciting.' And my dad who was the shoemaker in a small mining village in the north of England said, 'You can be anything you want to be, you know.'

"I probably would have been a poor astronaut," Ann muses, "because I'm actually quite claustrophobic. When I saw the lunar landing module years later at the Smithsonian I thought, 'I wouldn't fancy being stuck in that,' but the point being about having role models. I don't think role models have to be in the working environment; they can be the first astronaut" – or the first female prime minister.

"Obviously I've made career changes in my life. I don't want anybody to walk away thinking I thought engineering was much better than banking. If I actually thought that, I would have gone back to engineering, wouldn't I?" she asks rhetorically. "I think that eventually finding what you love to do in life is something that connects you to your earlier life and your skills.

"Where I ended up is very much where I started, in that the engineering and the technical and maths side are combined with an understanding of how money flows, so for me this is the perfect job."

♀ ♀ ♀

7	**In Memoriam** **Sarah Helen Darnley (1968 – 2013)**

In August of 2013, Sarah Darnley[23] of Scotland was among the 30% of female offshore workers employed by the catering industry and, by all accounts, loved her job as a steward.

Although she had worked off and on in the offshore catering industry for 17 years, and had been with her employer, Sodexo UK Remote Sites for much of that time, Sarah only recently had begun working a rotation schedule onboard the Borgsten Dolphin rig, which is operated by French oil giant, Total, and located in the North Sea.

Tragically underscoring one of the rare but real hazards of working offshore, Sarah and three male colleagues – George Allison, Gary McCrossan, and Duncan Munro – lost their lives on August 23, 2013, when the helicopter in which they were flying back to shore crashed into the North Sea just two nautical miles from land, making Sarah the first woman to be killed in the history of the UK's oil and gas industry.

At the time of the accident, Sarah's younger sister, Angela, was quoted in the media as saying, "Sarah loved her work. She never expressed any fears about traveling in helicopters despite the dreadful previous incidents."

Sarah and Angela were born in Elgin, Moray, in Scotland, just 13 months apart to Edmund and Anne Darnley. Growing up, Sarah took part in numerous school activities, showing an interest in the arts, starting in primary school where she honed her musical skills on the recorder. Later, in high school, she developed her artistic talents, while

---

[23] Sarah Helen Darnley: born March 3, 1968, died August 23, 2013.

also exhibiting athletic ability, particularly in gymnastics, a sport in which she won many trophies. Although she worked part-time while she was in school, eventually going to work for Asda,[24] Angela says Sarah "had no clear pathway at school."

After leaving school at 17, Sarah enrolled at Moray College of Further Education where she undertook a course in business. Even after completing her studies, she was unsure what career to pursue and continued working at Asda, transferring from Elgin to Aberdeen. Despite working full-time, she discovered the cost of living in Aberdeen was quite high and made the decision to go offshore. She found employment as a catering steward, which involved performing routine domestic chores, such as cabin service, laundry, and assisting in the kitchen. Although there were not a lot of women working offshore when Sarah started her career mid-90s, her family says she did not face any discrimination; however, Sarah did say she felt that it was harder to climb the career ladder as a woman.

Eventually tiring of life offshore, Sarah made brief forays into insurance and banking, typically working on a commission basis. Her mother Anne says Sarah enjoyed the challenge of a change and did well, earning merit awards on her examinations, but the pressure of possible takeovers and a downturn in the economy led her to resign. Angela also points out the income was less than what Sarah was accustomed to and, as someone who appreciated being able to travel and afford nice things, once again she was drawn to offshore work. In 2002, she returned to work for her previous employer, Universal, which had become Universal Sodexho,[25] and would remain with them until her passing (no byline, 2013).

Angela says Sarah was at her happiest working on the Leiv Eiriksson rig in the Norwegian sector, but eventually that assignment came to an end, and Anne says Sarah particularly felt the loss of the camaraderie that had been established. While on a tow from Norway to Turkey,

---

[24] Subsidiary of the US retail store Walmart. Source: Walmart Corporate
[25] In 1999, Kelvin and Universal Services consolidated their North Sea operations, forming Universal Kelvin. One year later, the name was changed to Universal Sodexho. Since 2008, the company has been called Sodexo (the "h" was dropped). Source: Sodexo Remote Sites

Sarah was offered employment there, but discovered that the local men didn't particularly like taking instruction from a woman and had a different work ethic, both of which put her off working there. However, as someone who loved to travel, she took advantage of the opportunity to visit many of Istanbul's historic sites and attractions before returning to the UK.

Sarah went back to work in the North Sea on an ad hoc basis, and also spent time working on the Falkland Islands, eventually securing a permanent position on the *Borgsten Dolphin* rig off the Shetland Islands, where Anne says she established the same level of friendship with her colleagues that she had had on the Leiv Eriksson.

By this time, she had spent many years on the water and, as Angela describes it, was "beginning to get fed up with the offshore life and the night shifts." Wanting to remain in the industry, Anne says Sarah had just completed "a pretty intensive" health and safety course in pursuit of a different role, although she would not have the opportunity to change career paths before the accident. "Sadly," Anne says, "that was not to be."

Dispelling stereotypes of women who work offshore, Sarah was nick-named "Lady Darnley" for her love of the finer things in life, which an unnamed friend, who was among the hundreds of mourners in attendance at her funeral service, was quoted in the press as saying included "the best cognac, Chanel, designer clothes, and first-class travel" (Nicolson, 2013). Angela confirms this and emphasizes, *"She enjoyed her life."*

An independent, single woman, Sarah never married or had children, although she was very close to Angela's son, Nicholas, and often talked to him about working offshore. After her death, some of the UK newspapers ran a photo of Sarah and Nicholas together when he was younger. Now 24, he had completed his survival course and was waiting to secure a job offshore at the time of the accident. Deeply affected, he said that he would never go offshore but a year later, following in his beloved Aunt Sarah's footsteps, he did and continues to work offshore in the North Sea as a handyman.

When Angela asked one of Sarah's friends why Sarah hadn't married, the friend laughed and responded it would had to have been someone very, very special to fill that role. Angela remembers Sarah often joking that she might have married a diver because they make good money or possibly someone that worked shifts opposite from hers. However, her family says Sarah was quite content with her life and her friends, many of whom are single, as well. Angela even goes so far as to say, "In Sarah's life there was no room for anyone other than her cat, Tilly." Her flat mate, Dustin Nell, remained in the flat and kept Tilly after Sarah's passing.

"She was never without a cat," Angela says. In Aberdeen, she first had a cat named Bella, but felt like she couldn't properly care for it while she was offshore, so she entrusted Bella to her parents. Then, "lost without a pet," Angela says Sarah got Tilly, naming her after Tillydrone, the area of Aberdeen where the cat had come from.

Animals were one of Sarah's great loves throughout her life. She had a horse named Charlie when she was younger and in 2012 she attended the Summer Olympics in London, where she was particularly interested in seeing the dressage competition. She had plans to go to the 2013 Horse of the Year show in Birmingham when she returned home from her hitch.

Travel was another one of Sarah's passions and she didn't mind traveling solo as she had friends all over the world, including London, Argentina, and South Africa. "Wherever she went, she would make friends," Angela says, and often she would arrange to meet someone she knew. For her 40th birthday in 2008, Sarah treated herself to a trip to the Dominican Republic. She had begun taking Spanish lessons in preparation for a holiday in Spain upon her return from offshore.

Sarah's immediate family describes her as "a popular person" among her colleagues and says she often would talk about her offshore family – "especially the challenges faced by those with family commitments," her mother Anne says – when she called home during her rota. The depth of their feelings for her was shown following her death by the number of cards and letters the Darnleys received from the people she worked with as well as their families.

At the annual Service of Remembrance, a new *Book of Remembrance*, which originally had been created by the UK Oil and Gas Chaplaincy to honor those who perished at sea, whether by accident or by natural causes, was dedicated in October 2013 to mark the 25th anniversary of the Piper Alpha tragedy. Sarah's name and those of her three male colleagues – George Allison, Gary McCrossan, and Duncan Munro – were read aloud and added to the new book at the service in the Kirk of St. Nicholas Uniting in Aberdeen led by the Reverend Gordon Craig, chaplain to the UK oil and gas industry.

One year later, a memorial video, which was shown onshore and off, was created jointly by Oil & Gas UK and Step Change in Safety to honor the four lives lost. Two years after the accident, Sarah's family attended the service that was held by the UK Oil and Gas Chaplaincy at the Sumburgh Airport Memorial in Shetland.

Sarah's mother Anne calls her "a one-off" and says, "Sarah was very much her own person and, yes, a free spirit. She did what she wanted when she wanted."

In a poignant tribute, pastoral assistant Paula Baker, who conducted the funeral service held in Elgin in celebration of Sarah's life, said, "[She] has left us as she lived – dramatically and on the crest of a wave" (Nicolson, 2013).

8	**Myrtle Dawes** **Chemical Engineer**

"I have more freedom than others who have chosen to conform
because now they have to stay within that straitjacket."

Twenty-five years after graduating from Imperial College in London with a Master of Engineering in chemical engineering, Myrtle Dawes has empirical evidence to support the assertion that young girls today are not quite sure what engineers do, just as she was not sure when her high school physical chemistry teacher, Dr. Croft, suggested during her second year of A-levels that she major in engineering at university.

"I said, 'Oh, what's that?'

"He just looked at me and said, 'You like physics and maths and chemistry. You'll be fine.'

"In those days, you didn't get any information on what engineering was," Myrtle says. "I thought an "engineer" fixed your car. Here in the UK we still use that term. So it wasn't very clear, but I'm very pleased I got that piece of advice."

Myrtle, the mother of a 13-year old daughter named Natasha, has witnessed firsthand the disconnect that young girls still have with engineering. A few years ago, she returned home to North Wales, having worked in the Aberdeen offices of Centrica Plc during the week. Natasha and her now 16-year old brother, Benjamin, live with their father, Myrtle's former partner, while she is away and she has them on weekends. She asked her daughter what she had done that week. Natasha said she had to write something in school about what her parents did for work, and Myrtle asked how she had responded. "She said, 'Daddy is an engineer.'

"And, indeed, he is," Myrtle confirms. "He does something similar to myself." She then asked her daughter what she said she did.

"She looked at me and said, 'I don't know what you do, Mummy.' I said, 'Well, I've told you so many times. I'm an engineer.'

"She cocked her head to the side and looked at me and said, 'What do you mean, like Daddy?'"

Photo Credit: Centrica

♀ ♀ ♀

Myrtle's experience as one of a small number of women attending Imperial College, known for its emphasis on science, technology, math, and now medicine, would be good preparation for her future career in the petroleum industry.

"There were very few women across the campus, in general. In fact, in Royal School of Mines, there were only two girls in the entire school during the time I was there. We had a few more in chemical engineering, even more in mechanical engineering, not so many in electrical engineering. But, certainly in mining, hardly any at all. Having said that, there was a stark lack of diversity on every level."

While she was at Imperial, she had two industry placements, as they were called, which unfortunately she says did not prepare her for the workplace in any real way. The first was conducting research at Cranfield University, studying slugs in large diameter pipelines.

As someone who loves traveling, the second placement in Zurich was "fantastic" and helped satisfy her wanderlust temporarily, but involved more research – in this case, gauging the surface tension of sewage water to determine the purity. "That sounds really gross," Myrtle acknowledges, but she reveled at the chance to work with innovative equipment under the direction of "an absolutely amazing Eastern European professor," who advised Myrtle that, even as a chemical engineer, she needed to know how to code. Ultimately, her research was the first paper she had published under her own name.

However, both placements were in very academic settings and not at all like the real world work environment, the reality of which, she says, "was a bit different."

Just as she decided to major in engineering more or less out of blissful ignorance – aside from the suggestion by her high school chemistry teacher – Myrtle stumbled into her first job out of university quite by mistake. She applied to British Petroleum (BP) Engineering to do process engineering and, instead, ended up being hired by BP Exploration. She accepted as a petroleum engineer, not knowing what the role entailed or what the practical application would be, only to find herself – someone with a fear of flying and a self-professed "poor swimmer" – assigned to a job offshore!

"I was a bit shocked to find out that I had to get into a helicopter to go offshore, but I had already committed. You drift into these things when you're that young without realizing what they mean."

She did the obligatory "five days of fire and five days of water, as we used to say," and managed to get through them due to the positive effect of "peer pressure.

"I just thought I had no choice. Everyone else was doing it and there were people who also didn't swim. Sometimes it's good to be in a situation with people who are exactly the same as you because they help you get through." She recalls being afraid to jump off one of the platforms. She and a male colleague, who also didn't swim, held hands and agreed to jump together on the count of three. One, two, three...

"He jumped and I didn't." The instructor then informed her no one was going to lunch until she jumped.

Having survived the safety training (something she still dreads, although she says she is a somewhat better swimmer now), Myrtle became one of the first black female engineers to work offshore in the North Sea. In fact, when she started working on the Forties Bravo, she was the only woman working on the facility (at least during her rota). There was, however, another woman working on Forties Alpha, the sister platform, and their paths would cross again later in their careers.

"In all that year (1991), I think BP took on eight graduates and four of us were women, but all of us went to work offshore," she says, although not on the same projects. "That was one of the first stream of graduates that was taken and it was quite a mixed group."

Her arrival offshore was viewed in a manner she calls "two-fold." She acknowledges it was "a bit of an intrusion for the way the men went about work at the time," but also found many of them were paternal in their interactions with her.

"I was reasonably young, as were most of the women going offshore, and most of the men were about our fathers' age, and so actually treated you very nicely and were really helpful." On the other hand, it was a very "undisciplined" environment in certain ways. Myrtle had been warned that "training videos" were streamed to the cabins at 9 o'clock at night, although she wasn't sure what that meant.

"So, of course, I tuned in," she says, laughing.

There was another incident, which she refers to as the "most embarrassing one for us all." She had been called to the OIM's office, but because the Forties Bravo was such a large facility, she had gotten lost. Walking into one of the rooms, she discovered about eight construction workers sitting amidst "wall-to-wall porn – the ceilings, the walls, everything." Not knowing what to do, she looked down, before getting the courage to eventually look up and admit, "I'm lost. Can someone take me to the OIM's office?" Needless to say, she suddenly had eight volunteers who were very eager to help.

As a result, about two months into the rota, Myrtle says the asset manager for the Forties informed the crew he didn't want any "training videos" streamed while there were women offshore, and that some of the "wallpaper" had to come down. A few months later, he then decided if the videos were not going to be streamed when there were women offshore, the videos weren't going to be streamed at all because of the complaints from the men who didn't get to watch them when they worked the same rota as Myrtle!

"He was very much disliked as a result," she says. "Things you just would not expect to happen today were happening then. It was strange; it did change how they operated. With the Internet, probably it's a little more private now."

As one of the few women – and the only woman of color – on her offshore postings, she says she didn't really have a mentor, and still doesn't. She resigned from BP after just a year and, looking back, says that is probably one of the reasons she left – not specifically because she didn't have a mentor – but because "I felt very lonely. Everything was so different from what I was used to and I just really wanted to go back to a place that was more familiar.

"I think that was the right decision for me at the time. I went on to do really exciting things. I'm not sure I would have done the same things at BP. I'd say I'm not really fond of working offshore. I'll do it, but it's not my preference. I managed to find a lot more work, which was interesting, that was not offshore."

Returning to London, Myrtle first went to work in safety, which was at the forefront of the industry following the Piper Alpha disaster a few years earlier in 1988. New legislation was being drafted and the industry was looking for engineers who had worked offshore and could help put together the safety cases and some of the quantitative risk analysis (QRA).

"I also worked on some other really fantastic things from a QRA perspective. We worked on the Hong Kong International Airport, and the Channel Tunnel QRA – most people don't realize there was a risk analysis for that – and the Jubilee Line extension of the London Underground."

After three years, Myrtle joined Genesis Oil and Gas Consultants as a senior engineer doing process engineering, "going back to perhaps what I was trained to do."

She then embarked on what would turn into a grand adventure working on a design contest for Terra Nova Energy Project to build the first ice-strengthened ship to be used as an oil and gas platform in the Grand Banks of Canada – "the same waters close to where the Titanic sank," a fact she mentions casually.

Hibernia, the previous platform, had field production costs of $7.9 billion[26], making it cost-prohibitive to re-create, so the decision was made to use an ice-strengthened ship. The project was an engineer's dream in that it was unique in terms of the amount of ice-strengthening that had to be done, and the fact that the ship was the biggest of its type at the time and featured a special turret, which could be disconnected, enabling the ship to sail away.

"We had subsea wells which we had to put in depressions – what we call 'glory-holes'[27] – we had to dig big holes and put the wells in! Otherwise, the icebergs would come and rip them up. So, really amazing kinds of hazards that you don't normally get in the North Seas or in the desert or indeed in Bangladesh," she says, referring to the Sangu onshore project on which she would later work. "This is when you start becoming really aware of hazards. You can't base them on what you've done before because every new project has something slightly unique."

All of the early, detailed engineering was done in London, during which time Myrtle gave birth to her first child. Terra Nova held her job open, although she told them she didn't know if she would want to return. After being home for three months, "I was so bored that when they phoned, I said okay!"

---

[26] Source: Subsea IQ Deepwater & Subsea Projects
[27] Large, open excavations.

Her infant son, Ben, in tow, she arrived in St. Johns, Newfoundland, and continued with the project for the next year. She was impressed by the support she received from the company, which she attributes in part to having a Norwegian boss, who would have been accustomed to Norway's generous childcare and parental benefits. She and her son were put up in a hotel, where a nanny had been arranged, until a house was ready for them. Although she was in her 30s at the time, she didn't drive, so taxis ferried her wherever she needed to go.

She points out that it was "remarkable" that the job was held open for her, as she was working on a contract basis. "Normally, they would never leave a job open for a contractor, but because of my knowledge of the project and what I brought to the table, they were very keen for me to continue. That pleased me."

While acknowledging that it was "a bit daunting going to work on my own with a small child in a foreign country," because she was treated so well and her contributions were clearly valued, Myrtle believed the arrangement would work – "and it did."

"All they wanted me to do was to work, so that was good. It was a long project; it takes time to design and construct these things. The construction was quite expansive. We had the turret built in Houston. The hull was built in Korea. It was one of the first times that what we call Floating Production Storage and Offloading (FPSO) was built in Korea at Daewoo. Parts were being built in the Middle East. We had the subsea systems being built in Norway. It was really an international project – a big mish-mash of nationalities – and a fantastic project to be on.

"Going to Newfoundland was great. The locals there are very, very different in terms of their outlook and expectations. Very big ownership of the project in St. John's. I used to get flyers coming through my letterbox about things we were doing on the project that people didn't like. In the UK, you wouldn't expect people to talk about your project, but they did there because they really identified this resource as belonging to them. I think it is a good thing. You find the same thing in Norway. They talk about Norway, Inc. a lot. There's a sense of 'I have to do the right thing because this is for me and upcoming generations.'"

Having finished the design competition for Terra Nova, Myrtle then, "for money reasons, no more," joined Halliburton Brown & Root on a project called Sangu,[28] which she says, "Was absolutely fantastic. It was the quickest project I've ever worked. From engineering all the way through to construction, it was literally only a few months. That's when I started to appreciate some of the more complex things in terms of culture and getting work done. It was the first time I got exposed to the fact that different countries take different approaches."

After Sangu, she would spend the next seven and a half years both onshore and offshore with BHP Billiton in a number of engineering roles – lead safety engineer, field development engineer, and project delivery manager.

Myrtle, who describes herself as "always having a good time," was once told she was "too outgoing for leadership roles," and that when the company was looking for future leadership, management was looking for someone like themselves because they knew that that person could lead. They didn't see that quality in her.

"I said that, obviously, I thought it was kind of unreasonable and showed a lack of diversity in terms of approach and style." While there was never any dispute about the quality of her work – something that was reiterated – it seemed to be her personality they felt wasn't suited to leadership.

"I made a decision that I was quite happy with who I was. That was fine, in fact, if that's how they wanted to run things. But I didn't necessarily want to change. I didn't want the stress of having to put on an act every day. And so I didn't, actually."

Eventually, she was promoted, although she felt she was never really given the opportunities she deserved nor was she able to fully contribute what she felt she had to offer.

"I left when my children were old enough for me to start working away from home, and it's been the best thing I've done," Myrtle says resolutely. "I'm not going to go and say they did the bad thing because it's worked out best for me. I turned it into a challenge. I still think there is an expectation about style and approach, which isn't very diverse and does exclude people."

Just recently, she was asked to take a survey that asked about "cracking the code" and whether women need to act differently in order to appear more confident. "I said I was quite happy to act just as I am. *We need to recognize that confidence looks different in people.* This

---

[28] Located in the Bay of Bengal, Sangu was Bangladesh's first offshore gas field development. Source: Cairn Energy

issue about putting ourselves forward... leaders have to be on the alert and be a little more objective about what they're looking for.

"I'm quite comfortable being me and, if anything, now that people have accepted me, I've even gotten worse!" she says, laughing uproariously. "I have more freedom than others who have chosen to conform because now they have to stay within that straitjacket."

From the beginning, it has always been important to Myrtle to have a career that fits her personality and she has found that in engineering. "I want to do something fun and interesting when I get up every day. I've actually found that's where I am. First of all, the job is not what everyone thinks it is. It's more a problem-solving job." A life-long learner with a passion and aptitude for maths, she keeps her skills sharp by continuing to take maths courses from Open University.

"When you're in a technological environment like we are, you can't stand still. You have to be asking, 'What's the latest technology, how does it work, what is the legislation, what's new, how do I need to do this differently?' We're in this constantly changing environment where you do everything that you could want, you're reasonably well-paid, you get to travel, you get to network for business reasons, and you meet such a diverse range of people.

"I think women *are* very good at developing these relationships. If you can use this to get work done, to support a team, to lead a team – and put aside some of the pride and follow from time to time – this industry is a fantastic place to be."

Since joining Centrica in 2009, Myrtle has assumed a wide range of roles, illustrating the diversity of opportunities that exist within the industry. She started in the Windsor office, working on a gas storage project, and then moved to Aberdeen to work with the engineering manager for a while. From there, she went to Norway for 18 months and worked with the project manager. "It was fantastic starting their management team from scratch, looking after five projects that had an estimated capex of a billion pounds each, and growing a team there." She then returned to Aberdeen in a different role as the project and decommissioning director.

"People expect it to be the same every day and it's not. Far from it. In fact, my manager wants us to be innovative and different and find new solutions, trying to be safer, cheaper, etc. It's just fantastic. To me, it's just not the job that sells you when you hear the term "engineering." I used to identify myself as a technical engineer. Engineering is something you use to do the job. It doesn't matter if I lead and manage, I use the same skills – the thinking, the problem-solving – I learned as an engineer."

Her manager just happens to be Colette Cohen,[29] now the senior vice president for the UK and Netherlands, and the female engineer who worked on the Forties Alpha, the sister platform to Forties Bravo, where Myrtle worked for BP at the start of her career.

"Colette is probably one of the most competent leaders I've ever had. She has a great ability to consume a lot of information. What do I find different about her to a male boss? It's not different in terms of how much she pushes us to get things done. But, what is different I think, because is she had to really struggle and do things differently in the industry to get to her level, she's very, very open to new things. If she asks me to do something and I say, 'That's not going to work because...' she'll challenge me on that. She doesn't accept the status quo and I think that's natural for her because, had she accepted the status quo on anything, she wouldn't be where she is at the moment. She's made a great, huge amount of change since she's joined the business. She's really a force to be reckoned with."

Myrtle's career has come full circle in other ways, too. She is, once again, going offshore regularly. "I did find a way to get out of that for many, many years," she says frankly, "but I think because of the big push on ownership and leadership, especially in this field from a safety perspective, making sure people know you're involved in the work, you really have to go offshore. I go the terminals as well, so that people can see you there, you can get to know everybody, you can see the work you're asking people to do, see the kind of conditions people are having to work in. It's become more important. Maybe 10, 15 years ago, I wouldn't have to go as often, but now we actually set ourselves objections in terms of what we call "boots on the ground" and making sure we're there and aware of what's going on."

This year marks her 25th in the industry and she has weathered more than one downturn in that time. "We always talk about how we want other people to believe in us and recognize our capability, but a very important thing is to believe in yourself and what you can offer because that keeps you grounded in the day-to-day of what you're doing rather than wondering about what's going on. I say frequently to my team, *'There's no one who's going to rescue us at the moment; it's all in our*

---

[29] Colette Cohen was named CEO of the Oil & Gas Technology Centre on July 14, 2016.

hands.' Every industry has some cycle. And those that don't, you probably don't want to stay there because it really is quite static."

As much as Myrtle enjoys working in oil and gas, it is her belief that many of the skills and expertise acquired and put into practice by those of us who are employed in the industry can be transferred to other industries, should that day come. Knowing that, she believes, should relieve some of the concern about being involved in a cyclical industry.

Even given the current downtown, she says, "I see that we are still recruiting. I know how many engineering graduates are in my team. That's why I think it's really important to make people a little bit more resilient to change. You've got to accept that there will be some change. I've seen change come for good and bad reasons and it's not hurt me at all. In fact, if you can always be part of the change, you'll probably continue and be part of what's going on."

Regarding the last point, she says, "That's really seen me through a lot, and that's why I like to keep my hand in, so that I can actually apply what I'm doing. As teams shrink, what I've seen here, for example, is I've been given more responsibility because I'm quite flexible, very good in first-time situations, and will take on that challenge."

Another challenge she's proven more than willing to take on is helping young people – girls, in particular – understand what engineering really is and the opportunities associated with it. She is an ardent supporter of projects like STEMNet and STEMettes, the latter of which was co-founded by Jacquelyn Guderley and CEO Anne-Marie Imafidon, MBE[30] – whom Myrtle describes as a "math genius, who really does energize the girls" – and has been involved with both organizations.

"The girls ask me, 'Isn't it a problem working in an office full of boys?' I say, 'It's not actually full of boys. There is probably an equal amount, but not necessarily an equal amount doing the technical work.' She has participated in events where groups of girls from local schools spend the day at Centrica, observing the work engineers actually do.

"It's really just to de-mystify it. There's an impression that it's dirty and you have to lift heavy things, like you're working in a coal mine. I think it still has that image. They can see we're at our desks, we're reasonably presented. You should see me at the moment," she says, laughing. "I have long nails. It's not an issue of becoming quite dull and grungy. *Then* you can start talking about the work and showing them

---

[30] At the age of 27, Anne-Marie was awarded a Member of the British Empire (MBE) in the 2017 New Years Honours for services to young women and STEM sectors.

some of the really interesting things that engineers actually do. That's the kind of thing I'm really quite passionate about."

That's understandable given that she has a daughter who is still "quite surprised" that her mum is an engineer. "You realize that it's not necessarily that I have ever said to her what an engineer looks like. But even growing up in a house with two engineers actively working – she's never known me not work – but because of all the other things she sees on TV, and all the things she reads, it's still difficult for her to accept that I am an engineer. That was startling to me. I spend time with her, obviously, making sure she knows that, of course, I am. That's a lesson for us all, really. Yes, that I could be doing something like her father. This is why I then realized all these other messages that the young girls get are quite strong. Whether if it's that they take them in a profound way and they sink in, but they get these messages and they're not necessarily what we want them to pick up."

Myrtle uses her own career as a case in point, noting how "tentative" and by chance the trajectory has been that has taken her to this point, starting with her course of study at university and her first job with BP. She feels those decisions are often "too tentative and not intentional enough for some of us and that's why it's easy for them not to happen, as well." She stresses the need to focus on girls, in particular, not only so they don't lose interest in STEM subjects, but also so they will feel confident that those fields of study will lead to a job, and that they will be taken seriously and seen equally.

These are things she talks about to both her 13-year old daughter, who wants to study medicine, and her 16-year old son, whom she says, "has the advantage of thinking, 'I can be an engineer' because he says he wants to be one."

She once said her Plan B is to become a maths tutor when she retires (Adom, 2014). In the meantime, if there is anyone for whom the word "engineer" still conjures the image of an auto mechanic, they need only take one look at the press photos of Myrtle accepting the 2015 MER[31] award on behalf of Centrica. Dressed in a platinum-colored, floor-length gown, surrounded by her tuxedoed male colleagues, she proves that engineering isn't always about operating pumps; sometimes, it's about wearing them, too.

♀ ♀ ♀

---

[31] Oil & Gas Authority Maximising Economic Recovery (MER) UK Award

In February 2019, after nearly a decade with Centrica, Myrtle Dawes decided to take a break and explore her next opportunity in global energy. She was last seen on a beach in Jamaica.

9	**Anne Grete Ellingsen**
	CEO, Global Center of Expertise, NODE

"Make yourself visible."

Anne Grete Ellingsen, the first female offshore platform manager (OIM) in the North Sea, might well view the 1985 documentary *Norwegian Women Offshore* (NPD & Statoil)[32] as somewhat of a personal time capsule. At the time of filming, Anne Grete, as OIM, was responsible for the Frigg drilling platforms on both the Norwegian and UK sides, and later for the commissioning of the *Heimdalfeltet* platform (commonly known as Heimdal).[33] Recognizing early on that there was a power dynamic at work, the (then) 32-year old petroleum engineer was determined not to be distracted by it and instead focused on the job she was hired to do. "Much depends on how you see yourself... and believing in yourself," she says in the film.

Born in Bergen, Norway, she was her paternal grandmother's first grandchild. Despite coming from a wealthy family, her grandmother's father was very conservative and, although she had top grades, she was not allowed to continue her schooling after the twelfth year. At that time, young women from "good families" were expected to marry early, keep house, and be proper hostesses. However, being widowed at just 35 years of age left her without the support of a husband or the benefit of higher education.

As a child, Anne Grete and her family lived with her grandmother for a period of time and her grandmother's influence would have a profound

---

[32] Norwegian Petroleum Directorate (NPD)

[33] Discovered in 1972, Heimdal is an offshore natural gas field in the North Sea located 180 kilometres (~112 miles) west-northwest of Stavanger, Norway. Source: Norsk Oljemusem

effect on her. She recalls slipping into her grandmother's bedroom every Sunday morning, knowing that *Farmor* kept chocolate hidden in her bedside drawer. Conspiring, as they shared the forbidden indulgence, they would talk about the future and her grandmother would tell her repeatedly, "Remember to get an education. Be independent. Get your education," something which she had been denied.

By the same token, Anne Grete's father "always treated my sister and me like the boys; there was no difference." Whether it was working on the boat or the house or their country place, he expected them to do things that were considered unusual for girls – starting the machines, as well as maintaining them, and building things. He also urged his daughters to pursue an education, which was a great source of encouragement to Anne Grete.

By the time she reached university, she had developed a real interest in mechanics and electronics, but, in those days, it was unusual for women to study engineering. Instead, she was advised to go into a "typical female area of study," such as chemistry, in order to pay her student loans and increase her chances of securing a good job.

Production from Norway's first major discovery, the Ekofisk field – to date, still one of the largest offshore oilfields ever found[34] – began in 1971, during the time Anne Grete was studying for her chemistry degree, and she recalls the excitement and attention surrounding Norway's burgeoning oil industry.

Despite misgivings that it was the right field for her, she earned a Bachelor of Science in chemistry from Bergen Regional College in 1975.

---

[34] Source: The Norwegian Oil and Gas Association

She then took a summer job at a hospital, where she discovered, "It was not at all like what we learned in the laboratory," and realized, "This is nothing for me."

Curious as to whether there were better positions in the emerging oil and gas industry, she went back to school for a year to study oil and refining and then applied for a job with the *Oljedirektoratet* (Norwegian Petroleum Directorate or NPD)[35] in Stavanger, the oil capital of Norway. She says she was lucky to get the position as a junior engineer, but as the only woman, "I think most of the guys thought I was going to make coffee and carry their papers, which I had no intention of doing," she says, laughing in amusement at the memory.

It was an interesting time because, as Anne Grete says, "It was the very starting point of the oil industry in Norway," and even the terminology had yet to be defined. When discussing safety, the expression "oilfield practice" was used, followed by the question, "What is that?"

"And no one could say exactly what that was." Anne Grete was part of the team given the task of developing safety regulations for the Norwegian Continental Shelf. She doesn't believe that, despite being the only woman, she took a more cautious approach or emphasized safety more than the men.

"No, I don't think I could make any distinction because we were working mainly with facts. The safety standards we were working on were how to keep the wells safe, what type of closure of valves, how many valves, what the closure time should be – it was more the technical side of safety, not the soft side. I don't think I could say I had a different view from my male colleagues."

As someone who was new to the industry, she was very excited to find herself working with experienced engineers from all over the world, participating in key discussions about safety, and having the opportunity to travel offshore to witness the operations.

Her boss, Magne Ognedal, sent a telex (she laughs at the memory – "we didn't have many computers") to Phillips Petroleum, the first operator in the North Sea, and the operator of Ekofisk, advising, "Mr. Ognedal and Ms. Ellingsen are coming out for a trip to the Ekofisk field."

---

[35] Governmental body for safety and resource management, reports to the Ministry of Petroleum and Energy (MPE). Source: Norwegian Petroleum Directorate (NPD)

Two days later, he received a telex, inquiring, "You mean *Mr.* Ellingsen?"

He replied, "No, I really mean *Ms.* Ellingsen."

And, again, came the incredulous response, "Are you *sure* this is right?"

Her boss said, "Yes, she is an engineer and she is going to work with the safety regulations, so it's absolutely necessary."

In 1976, the Storting (the Norwegian Parliament) had adopted The Working Environment Act,[36] which states that the working environment should facilitate adaptations for both sexes and the geographic scope includes "fixed and mobile installations on the Norwegian Continental Shelf."[37] The NPD had received a memo from the Ministry advising that this also should be adapted to the offshore sector.

Phillips Petroleum contacted the Ministry and asked if it was necessary for Anne Grete to go offshore. The Ministry, in turn, asked her boss if it was essential for her to accompany him offshore. Her boss responded that it was, and then asked if the Ministry would like him to rip up the memo he had received about preparing for both sexes in the industry. The response from the Ministry was, no, of course not. Her boss then concluded it was absolutely necessary for Anne Grete to go to the Ekofisk field.

At the time, there were no separate shower or toilet facilities for women, so she had to use the platform manager's cabin. Prior to her arrival, the men had been informed that if they were to say or do anything inappropriate, the consequences would be severe. As a result, Anne Grete says, "I was very lonely out there," but she still views it as a positive experience.

"Seeing things in action is something you have to do when you're going to be part of a team making safety and operational regulations, so it was absolutely necessary for me to be there. After that, I frequently went offshore to both the Ekofisk and also to the Statfjord,[38] which came along a bit later."

An unexpected surprise was meeting first- and second-generation Norwegian-Americans, many of whom had Norwegian names and were looking for their families. Anne Grete recalls a man named Ron Afdahl,

---

[36] The Working Environment Act Section 1:1c. Source: The Norwegian Labour Inspection Authority

[37] Source: Government of Norway

[38] Discovered in 1974, Statfjord is the largest oilfield in the Norwegian Continental Shelf (NCS). Source: Statoil

the start-up captain for the Ekofisk Center, who had a fascination with Norwegian stave churches from the Viking period and, along with his wife, was collecting data on the churches. During the offshore period, he actually spent his free time making model churches from matchsticks.

"Before he went home, he was traveling around Norway giving lectures on Norwegian stave churches in a real Oklahoma dialect," Anne Grete says, laughing. "He really was an expert."

Soon after joining in the Petroleum Directorate, she discovered a bachelor's degree was not enough; she needed a master's. At the time, no Norwegian university offered a master's degree in petroleum engineering, but the Stavanger Regional College had set up "sandwich courses" together with the University of Houston (USA), enabling students to attend school for two weeks to take master courses in petroleum engineering, and then return to work for six weeks.

"The first four years I was working, I didn't have much of a holiday; I was taking my master's. But that was very interesting. I thought that to have a master's degree – especially in an American environment, a French environment [where she would later work] – was important."

Another important discovery Anne Grete made early in her career was how to be a specialist. She remembers sitting in a meeting with the top management of Phillip's Petroleum and the specialist engineers from Phillip's, along with a man from an agency that was supporting them. The discussion centered on the flaring of natural gas and the damage to the flares, particularly in the Ekofisk area, from the perilous wind conditions in the North Sea.

"I was amazed at the knowledge of this guy. He knew the standards for designing equipment for flaring gas issued by the American Petroleum Institute (API) 520 and 521 and would quote from them. It was the first summer I was working at the Directorate, and I decided I was going to learn this by heart," she says, laughing. "I was using a lot of time to read these standards and I think I understood them quite well, although not all of them, of course."

Later, during another meeting, Anne Grete realized the man had misquoted one of the standards. "I said to him, 'I don't think that quote is like that, it's like this,' and he turned around to me and said, 'Are you an expert?' and I said, 'I guess I am.' So, from then on, I was an expert on flaring!" she says, laughing at her own daring.

There was also the highly-experienced machine engineer she had to work with, who did not appreciate having to go offshore with a novice – and a woman, at that – but was forced to do so. He expressed his displeasure in no uncertain terms by instructing her not to ask him questions in front of the other specialists, but to wait and consult him privately. Instead of being offended, Anne Grete found a way to turn this to her advantage.

Because the helicopters flew to different platforms, she and the machine engineer spent many hours sitting together in the helicopters, during which time she says, laughing mischievously, they developed a "special relationship." As he required frequent toilet breaks, her job was to keep the helicopter on the platform until he was onboard again. "I said to him, 'I'm going to keep the helicopter for you, but I have three questions you need to answer before you go.'" Despite his initial irritability, he was such an authority that Anne Grete considered it a "pleasure" to listen to his explanations.

On April 22, 1977, just two years into her career, Norway would experience the very thing she and her team were developing safety measures to prevent when a blowout occurred on the Ekofisk Bravo production platform. Fortunately, all 112 crew members were safely evacuated, but it was the first major uncontrolled blowout on the Norwegian Continental Shelf and the industry was shaken.

Anne Grete was not offshore at the time, but she vividly remembers Gro Harlem Brundtland, then Minister of Environment and later Norway's first female prime minister, going onsite. "There were a lot of Americans there, there was a special team for handling blowouts, there were all these guys from the media, CNN was there – all men. She came into the room – and she's not very high, one meter, 60 cm[39] or something like that – and she took command completely. I remember, I was so impressed by her," she says, the admiration still evident in her voice. "She was also very inspiring – the fact that it was possible to have that sort of authority, just coming into the room and taking control. *That* was impressive. She really did a splendid job explaining the situation to the world, and in fluent English. It was amazing."

After four years with the Directorate, Anne Grete joined France's Elf Aquitaine[40] in 1980. It was her responsibility to get the platform on the Frigg field located on the Norwegian and UK sectors certified and ready for start up, which involved constant communications with the UK and Norwegian authorities.

---

[39] One meter equals 3.28084 feet (~5' 2 ¼ ")

[40] Acquired by Totalfina in 2000 and called TotalfinaElf, the company now known as Total SA was renamed in 2003. Source: Total SA

After a year, Elf suggested she start offshore on a full-time rotation schedule. She first went to the south of France, where the Lacq gas field was discovered in 1950, to learn more about drilling and actually participate in onshore drilling.

"I was second in command on a drilling team that was traveling between Pau and Bordeaux, drilling wells in reservoirs with a cap rock of salt, where there were both old gas and also new wells, to improve the recovery rates."

With valuable drilling experience added to her CV, she went back to Norway in 1981 and joined operations offshore in the Frigg field as the platform manager/offshore installation manager (OIM) and the first female OIM in the North Sea. She was put in charge of the drilling platforms in the Frigg field, which is partly on the UK side and partly on the Norwegian side. "Working half a year on each side was very good for the tax regime at the time," she says laughing, and had everyone vying for that schedule.

It is also where she observed some of the cultural differences between Norway and the UK, where she says there is much more of a hierarchy. "In Norway, we normally solve things through dialogues. If a union representative acts outside the frame[work] in the regulations, you can always go back to the central union and have a discussion how to adapt. In the UK, you have to use much more verbal force. They are testing you to see if you really want to go through with what you say."

As an example, she cites an incident where there was a large boat laying aside the platform and she noticed the crane driver was not in the crane. She found him in the cinema watching a film. "I said to him, 'You know, there is a boat alongside that needs to be offloaded.'"

He responded that he was off-duty.

"I told him, 'Your duty is to offload the ship and if I don't see you in the crane within ten minutes, I'll have to send you onshore and you will lose your job.' It's that kind of threats both ways, which is different from the Norwegian attitude. After a few of these exercises, they understood I was serious."

She wasn't only working with men, though. In addition to women nurses and catering stewards, the first female operators were hired during her time as platform manager on the Norwegian side.

"My boss, who was French, was a bit shocked that I would like to hire female operators. That was the first step. Now we see a lot of female operators offshore. But on the UK side, it's much more difficult." In her experience, "The English are very conservative in comparison.

Norway has had women on ships for many years. There were very few on the UK sector. On the UK side, it was normally me and the female nurse."

Her work completed on Frigg, Anne Grete embarked on the required water safety refresher training in preparation for transferring to Heimdal as commissioning manager.

She calls smoke diving blindfolded, passing through barriers within the container, the intense heat, and, of course, the infamous helicopter escape – "a pretty tough test" – but there was an even worse experience yet to come.

On a "nice, sunny day in January with a very cold wind blowing" and an air temperature of minus 10C (14F), she, eight men, and one other woman were taken by lifeboat to a *fjord* in Stavanger and left to "float like corks" in the water in their survival suits. They were assured the helicopter would be there to pick them up in five to ten minutes.

One hour, two hours, two and half hours passed, and still no sign of the helicopter. There had been a miscommunication with the boat. Even with a survival suit, in water that is ~4-5C (40-41F), "you start losing feeling in your feet and your fingers." Of course, without it, "you wouldn't last long in that type of conditions."

"When the helicopter finally came and let down the ladder, the other girl and I went toward it, calculating we had some gentlemen. But we didn't. We were the last ones on board. They really stepped on us. We had to put on blankets to heat up again. So that was my ultimate test of gentlemen – and they failed! Absolutely. Those guys knew, if I was offshore, I was a manager. I worked with all of those men and I reminded them every so often that I was the last one onboard!" Anne Grete says, laughing.

She then assumed her role as commissioning manager on Heimdal, ensuring that the equipment was installed properly and in working order before going into the start-up phase, a very intense time she compares to having a baby.

"It's a nine-month period. You live offshore two weeks on/two off. At that time, we were doing all the hook-up and operations offshore. Today you would do more of that onshore."

Heimdal was like an offshore United Nations with a team of 50 to 200 people from 30 or 40 nations, working 12 hours a day "plus, plus," she says, bringing in the project on time and on budget, something she has said she considers one of the greatest achievements of her career.

Having such an international team led to some creative cultural exchanges. "We celebrated Christmas and we also had a New Year's party, where we had a procession and awards for the best costumes. It was really a great party," Anne Grete laughs knowingly. "You have to make it fun; otherwise, you go a bit crazy."

Every Monday between seven and nine in the evening, nominations were held for the Fork's Club, which was taken quite seriously. Members had to qualify and membership was designated by a fork made of light wood on their overalls. The challenge was to put together a meal that was "a bit bizarre." Every meeting ended with coffee *avec*, which on a Norwegian menu refers to coffee with Cognac. "Offshore it was coffee with fried eggs!"

Heimdal was finished and it was time for a new project. In fact, she took on two – she became pregnant with her first and only child, a daughter, and went onshore to act as Elf's liaison with the contractors who were helping build the North East Frigg field, Norway's first subsea completion system.

Her husband, also an engineer, had returned to Norway from an assignment in Gabon, and they found themselves working in the same department in the same company. At that time, no one was aware of that they were a couple because Anne Grete knew she would be the losing partner, if they were thinking about long-term career advancement. She and her husband decided that she would go onshore and have the baby, and then change positions. That decision was made all the easier after she returned from her four and a half months of paid maternity leave only to be passed over for a promotion in favor of a trainee from another company that worked in cooperation with Elf in the training of personnel.

She recalls the fury that she felt and says she left because her competence was not valued. Looking back, she was quoted in a 2012 article in *Rigzone* as saying, "I think that's one of the patterns you'll see from women in those days. You moved sideways instead of up when you had reached a certain management level" (Kammerzell, 2012).

She then joined Statoil in a managerial role for downstream, a different sector of the industry for her, that involved taking products from the refinery into the market and also the distribution of gas oil to the platforms. At the time, Statoil bought Exxon's activities in Denmark and Sweden, and British Petroleum's activities in Norway were merged, with Anne Grete helming her group's development of Statoil's operations for the downstream oil and gas market.

In 1990, three years into her seven-year tenure with Statoil, she began serving a two-year term as Deputy Minister of Petroleum and Energy. While the Minister is accountable to the Storting and has more of an outgoing role, the deputy minister is more of an internal position focused on what is happening within the energy sector. For Anne Grete, it provided the opportunity to see the industry from a "bird's eye point of view."

"The oil prices were very low – $10 a barrel. There was a lot of discussion that Norway should also go into Oil Producing & Exporting Countries (OPEC). We had meetings with the Saudis because they wanted us to decrease the operation and join OPEC to have sort of a common agreement on prices, but that was not part of the Norwegian policy. It gave me a very good perspective of the dynamics of the industry and how closely the energy sector is linked to politics and global security."

As she explains, the Ministry is responsible not only for the oil and gas sectors, but also for the renewable sector. At that time, the Ministry was in charge of deregulation of the hydro-power sector in Norway, the first nation in the world to do so. "We were actually pioneering the separation of the net from the production as an area of competition with market-based prices for electricity. The net, as a natural monopoly, where all the producers should have access at pre-determined tariffs, and the production, as the part of the company that was competing with market-based prices." Once again, she was at the forefront of change within the industry. Later, she points out, a lot of other countries, as well as the European Union (EU), would follow Norway's lead in restructuring the electricity energy sector.

Once she had fulfilled her two-year term as deputy minister, Anne Grete rejoined Statoil as part of its management group in Stavanger. Although Norway had yet to put its quota system in place – that would come later with the passage of the gender representation act[41] in 2004 – she achieved the position of company director.

She left Statoil in 1994 after seven years and, with an enviable CV and experience in both the upstream and downstream sectors, would go on to sit at the top of a number of boards, serving as chairman of Baltic Hydroenergy AS, Dolphin Group ASA, Havsul I, and Interoil Exploration & Production ASA, in addition to being a non-executive board member of several listed and non-listed companies.

---

[41] Source: Government of Norway

Having been on both sides, Anne Grete says, "It's hard to find women with good operational experience. Many women work in the softer side in economics and human resources (HR). To find top executives for boards is not that easy, unfortunately."

She supports Norway's quota system requiring 40% female representation on the boards of public limited companies, but says the job now is to get the boards to focus on making sure there are more women in top management in listed companies.

While she appreciates increased attention to the issue – Norwegian newspapers regularly name companies that have no women in top management – she says, "Often, when I have been sitting in top management groups, we have been asked to find candidates for promotions or management programs. Perhaps I'm the only woman in the group – or we may be two or three – and very often, we are the only ones that are coming up with a list of women to be promoted. *The network for many men is [other] men.* Searching for qualified women takes time and many men don't put priority on that."

Anne Grete then stepped away from the industry and she and her family lived in Oslo for 15 years, where she assumed various top management positions before moving to Kristiansand in the southern part of Norway, where they live now. She took a business development role with Agder Energy, a renewable energy company, looking for start-up investments, particularly in Africa, South America, and the Far East. Once they had established a few projects, the board decided that they were too costly to develop and it only wanted a minority share in the projects. Anne Grete was asked to become the manager of the hydroelectric water production for the region but says the role was not the right fit for her.

"I like projects. So I decided to start my own business, Energy & Management AS, and I was hired as a manager for Vestavind Offshore in charge of the first offshore wind park concession in Norwegian waters."

The project was in the planning phase, but because the cost of offshore wind projects makes it difficult to obtain financing, Anne Grete was tasked with determining what best practices, whether in technology, knowledge, competence or procedures, could be derived from offshore projects and applied to the wind project in an attempt to decrease costs. Again, not enough capital was raised to finance the project and it was put on hold. The low price of gas, which, in turn, affects the price of electricity, Anne Grete says, "makes it difficult to lift the renewable project."

Two years ago, she returned to the oil and gas industry, joining Global Center of Expertise (GCE) NODE, a private-public partnership, which is a part of the Norwegian cluster program. Its purpose is to enhance collaboration within research and development, and competence building between, in this case, companies in the global energy and maritime industries.

At the age of 60, "I started a new job!" she says brightly and launches into an explanation of the three levels: Arena, "immature" clusters in the early phase of collaboration; the Norwegian Center of Expertise (NCE), which receives more grants from the industry, and has started to coordinate research, development, and competence programs; and the highest level, which is the Global Center of Expertise (GCE).

"GCE NODE, which I'm heading now, was the first to gain that Global Center of Expertise status in 2014. We are also part of a European Union (EU) cluster program, where we were one of the first Norwegian clusters with Gold status and one of the top ten in the cluster program in the EU after the re-certification this summer. Seventy-five companies participate in NODE, some of which are part of large international listed companies, like Cameron and National Oilwell Varco. The production of drilling packages is done in Agder, and the companies located here have between 75 – 80% of the world's drilling packages. We also have a lot of companies that are leading in hoisting equipment and also mooring, shipping, and platform solutions. We have been delivering modules to China, which is impressive, I think."

She is CEO of an eight-person management team that is made up equally of men and women; however, there is not a single woman on what is literally an eight-*man* board. Because it is not a public limited company, it is not bound by the gender representation act. Within the cluster, there are two networking groups, NODE Eyde Women (NEW), which offers support to professional women in "hard-hat jobs" and NODE Young Professionals, which is inclusive of both women and men.

Anne Grete feels, "A lot of women are very focused on their profession – be a very good engineer – but they should also look at the hierarchy and how decisions are made and how they can get their decision to be the right one by using the dynamics of the hierarchy." She also advises her own daughter, now 30, and an engineer ("surprise, surprise"), and other women she mentors, "Make yourself visible. Write articles, participate in the industry, give lectures."

She was surprised when her daughter Siv, who is a lean expert and already heading her own team, didn't go into the energy industry, although she still holds out hope.

"First, she wanted to be a princess, and we made all the calculations what it cost to be a princess with the castle and dresses and everything. She decided that she had to earn a lot of money to be able to finance that. So then she changed to a chef because we're very fond of making food. During weekends, that's a project we have in the family. We always try to make nice meals, experiment a bit. So we said, that's fine, but you have to work in some of the very best restaurants in France and then you will be treated like a slave in order to have the best resume and the best CV, but we will support you if you want to do that. After thinking, she came back and said she wanted to be a lawyer. At the time, there were all these American television series about lawyers. We thought, okay, that's fine. All of a sudden she came and said, 'They accepted me at the Norwegian University of Science and Technology in Trondheim.' We were really surprised because that was the last thing we expected. She has a master's in mechanical engineering. She's a lean expert, so she's optimizing processes, something we can certainly use in the energy industry!

"When I left, the price of oil was $10 a barrel. The fun was gone because they started to standardize and there were not big projects anymore." Now there are new projects, but in the present downturn with oil slipping below $30 a barrel, "We have to do things differently. We have a lot to learn. It's still a very conservative industry, but there is a lot of technology we can move from other industries into ours to make it a lot more cost-effective. Absolutely.

"I'm following the agreement between Iran and the United States very closely and how that will affect the power balance in the Middle East and the energy market. How it will shift oil prices when Iranian production comes back into the market is one of the most interesting questions at the time now in industry," Anne Grete says, sounding like the Deputy Minister of Petroleum and Energy she once was.

"I'm back, and I love it!"

♀ ♀ ♀

In 2017, Anne Grete was inducted in the Offshore Energy Center Hall of Fame in Houston, Texas.

<table>
<tr><td>

10

</td><td>

Arlete Fastudo
Sonangol's First Female Marine Engineer

</td></tr>
</table>

**"I'm trying to make the way special for the women that
stay a long time at sea, away from their families."**

It was her late mother, Amália, a modiste, who "taught me to be an
independent woman, to work to get whatever I want, and to study to
achieve everything with success," says Arlete Fastudo.

But growing up she recalls aspiring to be like her father, Narciso, a
multilingual electronics engineer, who holds two master's degrees,
including one in foreign relations, whom she calls her "big inspiration."

"I really admire him because of his character and intelligence. When I
was a child, I used to say I wanted to be like Dad because I was so proud
to be his daughter in a country where it is difficult to get educated and
he made a way to give us the best."

Born and raised in Luanda, the capital of Angola, Arlete is the second
of five children and the eldest daughter. Although her father is an electronics engineer and her older brother, Adalberto, is an IT engineer, she
says, "Growing up, I never thought that I would be a marine engineer."
Instead, she aspired to be a lawyer. That changed when she entered high
school and was exposed to different professions. As she explains, "Our
high school is a little different from the American system or others. You
can learn technical skills, especially in the engineering field, but also in
information systems, telecommunications or nursing, and go to work
and start as a technician. I studied electronic communications and was
fascinated with this field. I was doing some training in one company, so
it was the practical side of what I was learning in school, to actually do
what I was studying." Students are given the opportunity to decide if
they want to continue with their chosen career path or explore another
field.

In 2009, Arlete was hired by Sonangol,[42] the Angolan state-owned oil company, and was one of the students awarded a scholarship by Sonangol for the marine program.

Photo Credit: Sonangol

She earned a high national diploma in mechanical engineering (HND-ME), attending the Academy of Education and Training (AMET) University in Chennai, Tamil Nadu, India, her first year and then the City of Glasgow College in Scotland her second year.[43] As part of the program, she earned her Officer of the Watch (OOW) license, which enables her to sail all over the world.

Traveling outside of Angola for the first time to attend university could have been a daunting prospect for a young girl who had lived at home with her family up to that point, but Arlete says she was more excited than anything. "I was proud of myself. I got a scholarship for the company to pay for my studies. I could finish school and not give more work to my father."

She also was fascinated with India having watched Bollywood movies on Angolan national television every Sunday. "I just imagined, wow, I'm going to India, I'm going to see all these music people. They are so beautiful; they sing and dance for everything. But real life was very different from the movies."

She would spend nearly two years in Chennai living in a hostel on campus at AMET University with two other Luandan girls, one of whom was recruited to study mechanical engineering and another to study petroleum engineering. Arlete says, "My English was not very good at that point and one of the girls helped me improve, so I am very thankful to her." In an industry that employs workers from all over the world, language skills are critical to avoid miscommunications and, as Arlete

---

[42] In June 2016, Isabel dos Santos was appointed the first female CEO in the company's 40-year history.

[43] On February 24, 2014, the Angola Maritime Training Centre (AMTC), owned by Sonangol EP and operated in conjunction with the City of Glasgow College, officially opened with the express purpose of Angolanizing Sonangol and the wider maritime industry in the region. Source: Sonangol's *Universo* magazine

notes, not everyone speaks English. In addition to her native Portuguese, she speaks English, and is conversational in Spanish with aspirations to become fluent given the opportunity to speak it more often. Currently, while onshore during her three months off work, she is taking French lessons at a local Luandan language school in order to broaden her ability to communicate while at sea and at various ports of call.

Once she had completed the first four months of study at AMET University, as well as "familiarization," which includes first aid, survival craft, and technique, she went out to sea for the following four or five months to put into practice what she had learned in the classroom, and then returned to school for nearly a year.

During familiarization, employees are given the opportunity to change their mind and leave the program to pursue another career path. Despite the frightening experience of exploring the inside of the exhaust manifold of a very large motor during familiarization, Arlete remained undeterred.

"I remember the first words of the captain. He said, 'Okay, girls on board, do you really want to do this? Is this really what you want for your life? If I were you, I would do something else. You can give up right now.'

"We said, no, we are here to sail and see how it goes."

He said, 'I have been captain for many years. If it's not easy for us men, think how it's going to be for you girls.'"

Rather than being offended, Arlete says, "I think he was saying, yeah, you can do it, but you have to show people you can. He was just joking, giving us a welcome, acting like a father. I was 19 and he looked at us like babies. It's a very lonely life, a hard life; I would not send my child out to sea." Having said that, she points out an interesting change in her personality that has developed over time. "You get used to it and then you do not want to be in crowded places. It affects your mind; you want to be in a calm place. Now, I prefer to be with my family or my friends," rather than going out. And when she is onboard the ship, she says, "It's amazing; you cannot stop admiring the sea."

Upon completing the first year of schooling in India, she went back home to Luanda for a period of time while her visa was being processed. She then moved to Scotland to complete her HND-ME through the program between Sonangol and Glasgow College, where she was fortunate to encounter a number of people who would be influential at the start of her career.

"I can't name only one mentor; it was not only one or two. I have trained and worked with many different people. What I did was get the maximum experience and knowledge of everyone I worked with and studied with during the process, and I'm very thankful to every single one of them."

Despite being proud of her accomplishments – earning a scholarship and finishing her schooling, then becoming Sonangol's first female marine engineer at the age of 23 – Arlete was somewhat taken aback by the expectations that were put upon her.

"When I talked to people, they made me feel like I was a role model. It was kind of heavy having that feeling, that responsibility, because people are looking at you as an inspiration and you don't even know it."

Her bewilderment is completely understandable when she explains that she has never met an experienced woman marine engineer, someone she could perhaps emulate or look to for guidance or simply talk to.

Then she brightens upon recalling, "I met a captain of a war ship and I was so fascinated with her. I asked her what it is like to be a captain on a warship and if she had been in a war, but she said, no, she had only done training and simulation." Perhaps the young marine engineer's next question was more telling. "I asked if she was married and she was divorced." She muses, "I think warships stay at sea longer."

That balancing act is something that obviously weighs on her mind as she looks toward the future. At 25, Arlete says, "I want to start my own family.[44] As a woman, it's very hard to deal with both the personal and professional side. I'm trying to make the way special for the women that stay a long time at sea, away from their families. Sea women must be allowed something regarding maternity leave." Because of the length of time spent at sea, "It may work for the guys, but when you work on an oil tanker or a liquid natural gas (LNG) ship, you cannot take a baby less than two years old. It's not the proper environment for a small baby or pregnant women." Once she has a family, she hopes to continue her career at sea. "If I could, I would." Again, she has no basis for comparison as the other marine engineers she has met have been young and single, starting out like she is.

For now, Arlete has her sights set on climbing the career ladder. Currently she is a 3rd engineer. "Once I feel confident in my skills, I will ask to go for examination for the next stage.

---

[44] The average age for first marriage for women in Angola is 21.4 years old. Source: United Nations, Department of Economic and Social Affairs, Population Division (2013). World Fertility Report 2012 (Report). United Nations Publication. p. Table A.1.

"I work on a crude oil tanker and while at sea the marine engineer must make sure that the ship runs without any problem. We do routine maintenance to prevent the systems and equipment from failing," and by that she means the hundreds of moving parts, machinery, and mechanical devices that must be kept in working order.

Depending on the route and the speed of the vessel, it can take up to 20 days for the tanker to reach its destination. A typical workday on the high seas is from 8 AM to 5 PM much like an average business day onshore, but that is where the similarities end. Arlete's duties encompass a myriad of responsibilities that range from starting and regulating the engines to the monitoring, maintenance, and repair of all vital systems – heating, refrigeration, ventilation, water, and electrical – to bunkering operations.

When her shift is over, she often spends time in her cabin listening to music or watching a film but, more often than not, studying to reach the next level of licensing. She then returns to the engine room at 9 PM for a "round," which entails checking the status of every single piece of equipment to ensure that it is functioning properly.

When the tanker is in port, the crew works in shifts: six hours on and six hours off. During that time, Arlete is in the engine control room performing cargo operations, including the discharge or loading of crude oil.

Given the long periods at sea, life can become monotonous at times. "When I am really involved in action, maintenance of equipment, such as generators, overhauling the systems, the time goes faster. Everyone knows what to do and how to do it, and that's when the time passes quickly." However, there is one task there is no way to improve. She confesses, "I hate working on the sanitation system."

Signing off for shore leave when the ship arrives in a port provides a welcome break. "I have been in so many countries I don't even remember the names of some of them, but the ones I remember are India, the US several times (Houston because our company has an office there and because of the port), the UK (London, Glasgow, Aberdeen), Spain, Brazil, Chile, Holland, and Mexico."

Her favorite port of call was the Caribbean island of Curacao. "I was only there a short time, but it was amazing. It was the first time I went off the ship and didn't do shopping. I visited the city, I saw the houses, and met people and tried to understand their culture. I got invited to a party. Their language mixes Spanish and Portuguese, so I could understand them."

As much as she enjoys the travel, Arlete says, "It's hard being out at sea for so long because you're away from home. Normally, when people go to work, they go home after seven or eight hours. When I did my first sailing as an engineer, I went for six and a half months, then I was supposed to go on vacation in China – I was buying equipment for the salon I own, Najema Beauty Shop – and I got the call to go home; my mom was sick. I got to the hospital the day before she passed away. I think she waited for me to get there."

Despite that devastating loss, Arlete persevered and returned to sea, even though she did not have the support of other women waiting for her. "Sometimes it's hard to be a woman in a man's environment because most of them are macho." When she joined Sonangol and discovered there were only two women in the Angolanization program, she wished she had more female colleagues. Although the number has now increased to 20, she says she has grown used to being in the minority.

With her slight frame, at one meter, 71 centimeters tall,[45] Arlete, who normally works alone, says, "If I need some strength sometimes, I get help from the motor men or an engineer, but we women think how can we use less manpower. We use our minds to make the work easier. You can minimize the manpower and the strength you need by using the equipment you have available." She recalls a male engineer she was sailing with whom she describes as "very tiny," who asked her rhetorically, "Do you think I can lift this motor? You don't have to apply a lot of strength; you can think [instead]. You will find a way." His comments stayed with her and she says ever since that day, "I have tried to work smarter, not harder, and find a better way to do things rather than using more force.

"I tell my boss, 'I'm the dirtiest engineer at work.' I'm dealing with the generator boilers and they're leaking like hell. I'm dealing with the black crude oil. I'm always washing my hands and sometimes I get an allergy from the powder because it's so strong. I keep a brush for my nails." Leave it to this woman entrepreneur – one who runs a beauty salon on the side – to come up with an ingenious strategy for dealing with the not-so-glamorous aspects of the job. "Sometimes, I wear black nail polish, so it's the same color as the oil."

As a marine engineer, Arlete keeps an unusual schedule. After being away at sea for four months, she then spends the next two months

---

[45] Slightly over 5'6"

onshore. If she is on the water for six months, she will spend the next three months on land. "When you go home, you have to do something; I don't feel comfortable doing nothing." She occupies her time with her salon, continually updating her skills with instructional videos on hair and makeup techniques; takes courses, such as the one in French, to improve her education; and studies to qualify as a second engineer, in order to reach her ultimate goal of becoming a chief engineer, before once again heading offshore to play an instrumental role in the running of a crude oil tanker. It should come as no surprise, then, that Fastudo translates from Portuguese, the official language of Angola, to English as "does everything" (*Univserso*, 2014).

11	**Abigail Ross Hopper** **Director, Bureau of Ocean Energy Management** **(BOEM) 2015 – 2017**

"[Growing up] I was told I could do whatever I wanted and that women were going to rule the world."

" My little sheltered world was cracked wide open." Having gone to prestigious Dartmouth College in Hanover, New Hampshire, to become an emergency room doctor, Abby Ross Hopper says she would be hard pressed to explain why during her sophomore year she trained to be a volunteer on a domestic violence and sexual assault hotline. "I became passionately interested in the challenges those folks face." She continued working on the hotline after she graduated as well as facilitating a support group for battered women. She then went to work at that same crisis intervention agency for the next few years mostly working with young teenagers, going into schools, providing education, leading support groups, and having to confront tough issues such as violent teen dating relationships, date rape, and child sexual abuse.

"Holy cow," she says in retrospect, "My world was opened."

Those experiences led her to enroll in law school convinced she would become a sex crimes prosecutor. Once she began the required course work, Abby says, "I kind of fell in love with contracts. I had a really fabulous contracts professor and I found it fascinating and decided that I wanted to do tax and corporate law." After graduating cum laude from the University of Maryland Francis King Carey School of Law, "I followed my heart into the corporate law world."

She went into private practice for nine years during which time she and her husband, Greg Hopper, also a lawyer, started their family, and in quick succession had three children.

"I had a four-year old, a three-year old, and a six-month old, and the thought of going back to private practice was a bit daunting." She had hoped to negotiate a part-time work schedule with her law firm where she had just made non-equity partner. A friend had recently become a general counsel and offered her a position as a deputy general counsel to Maryland's utilities regulator.

Thinking it might be slightly more balanced than going back to private practice on a part-time basis, Abby says, "I decided to take a jump and go into government practice." Despite having "absolutely no background in energy when I went into that job," she believes life and circumstances took her there. She remained in that position for almost two and a half years with no way of knowing at the time that it would be a pivotal step in leading her to the position she holds today.

She then went to work for Maryland Governor Martin O'Malley as his energy advisor.[46] "He was very interested in the potential of offshore wind off the coast of Maryland as a homegrown locally-sourced renewable energy option." Abby led a multi-year effort both on the policy development and then on the passage of a law that would allow an offshore wind farm to be built off the coast of Maryland.

"We lost twice before we won. I lobbied for him, I testified with him, I came here to BOEM several times and advocated on the federal leasing side. I had a lot of opportunities to get to know the industry, the players, the issues, and then articulate in a very, very public way why it was the right choice for Maryland."

Through that experience, Abby gained a reputation as an authority on wind energy, although she says she would never refer to herself as such. "Maybe this is where my gender comes through. I would never characterize myself as an expert; it makes me uncomfortable." Instead, she prefers to say she has a "strong background" in wind energy.

"I think a sustainable energy portfolio for our country is going to include lots of different technologies, but I believe *strongly* that offshore wind is going to be an important part of it. I think the resource is way

---

[46] Martin O'Malley was Governor of the US state of Maryland from 2007 – 2015.

too valuable to just let it sit offshore and not be utilized." She points out that it is not a big technology risk and that Europe has been harnessing wind energy for 25 years, while acknowledging that the US energy markets function differently than they do in Europe where they are more centrally organized.

"There's sort of a financing uncertainty in our markets here. I think when we crack the nut on how the markets and the technology can link up, it will take off."

Abby also believes the renewable energy space as a whole is a growing opportunity for women and girls. "The skill sets are the same as they are for other energy systems – engineering and more engineering. The environmental assessments around where you place wind turbines and solar panels require lots of environmental engineering, and understanding what animals and habitats are there, and how you're impacting the eco-systems. So there is a wide variety of career paths, but I do think that field is ripe for growth."

Knowing that Governor O'Malley's term in office would soon be coming to an end, Abby began applying for the position of director of BOEM, which had only had one director – a man[47] – since its inception in May of 2010.[48]

"I was hopeful that it would happen," she says of being appointed to the position in January of 2015. "I was excited and a little bit intimidated and glad for the opportunity. It was an honor; it continues to be honor."

Even though she is the first woman director, she doesn't feel like gender entered into the selection process and, in fact, calls her lack of background in the oil and gas industry more of a "defining characteristic." However, she believes there are two traits she possesses that helped compensate for that.

"I do have a strong background in renewable energy and offshore wind specifically and that's part of our portfolio here, so I think it sent a signal that it's going to continue to be part of our portfolio. I'd also like to think that it's based on my proven track record of my ability to grasp complex issues about which I'm not totally familiar. Every job I've had

---

[47] Tommy P. Beaudreau, the first Director of BOEM, served from June 2010 to May 2014. Source: BOEM

[48] On May 19, 2010, BOEM was formed as a result of a Secretarial Order by then-Secretary of the Interior Ken Salazar which divided the Minerals Management Service (MMS) into three independent entities: BOEM, BSEE, and The Office of Natural Resources Revenue. Source: BOEM

in my career, I've jumped into an area that I'm not an expert in." Abby believes Secretary of the Interior, Sally Jewel, who started her career as a petroleum engineer and former Mobil Corp. employee, felt confident she would be able to do that again.[49]

"As a political appointee, my job is to try to see the big picture and put the pieces together, how to envision and implement it." While she doesn't feel she has to be an expert on every aspect of BOEM's work, it is essential for her to have a strong foundation in the overall mission and be able to communicate that effectively. In order to do that, she relies on those who are the experts – and takes advantage of their knowledge and expertise to build upon her own.

"I'm not shy about asking questions; I'm not shy about what I don't know. I spend a lot of time reading and trying to synthesize information and figure out the connections between things."

However, Abby is selective about whom she chooses to ask those questions. "I ask them of people I trust. I don't necessarily ask a lot of questions in a room where I don't know the audience. It's an internal gut check whether this is a safe audience to ask a lot questions. But certainly with my staff and the people around me, I'm very comfortable asking a lot of questions."

Part of her job also entails answering other people's questions. "I've met with most of the major players in the oil and gas industry. I've met with them here in D.C., Houston, New Orleans, Anchorage, in Barrow, a remote Alaskan village... It's new and different to get out in the field and meet people on their own turf."

BOEM is responsible for energy resource development on the 1.7 billion acres of the Outer Continental Shelf (OCS), which essentially can be broken into two components: one is the environmental studies and assessment that analyze the impact of either oil and gas or wind energy development on the OCS; and "the other side of the house," as Abby refers to it, which leases the OCS for the purpose of either wind energy or oil and gas exploration.

Typically, Abby tries to set a couple of priorities and if something doesn't fall under one of those categories, she will delegate it to someone else, so that she can stay laser-focused on what she wants to accomplish.

---

[49] Jewel started her career working for Mobil Corp. in the oil and gas fields of Oklahoma as well as its exploration and production (E & P) office in Colorado. She later spent 19 years in commercial banking, where she began as an "energy and natural resources expert." Source: US Dept. of the Interior

She frequently finds herself in meetings, which don't always feel like "the work," but says, "I've come to peace with the fact that a lot of what I do is meetings and that *is* where a lot of the work happens." Her role also involves traveling a third of her time and being on the ground meeting people through BOEM's three regional offices – Alaska, the Gulf of Mexico, and the Pacific – in addition to another HQ in Sterling, Virginia.

"It's important to me to make sure we have good communication with those communities. I also spend a fair amount of time interfacing with external stakeholders – the industry, environmental groups, state and local officials."

While her position does not require her to go offshore, Abby has done so anyway to get a better feel for the resources over which her agency presides and to meet some of the people that work in the offshore industry.

"I have been offshore in the Gulf of Mexico (GoM) on both a production platform and drilling rig. BOEM doesn't inspect those installations; Bureau of Safety and Environmental Enforcement (BSEE), our sister agency, does that but I thought it was important to get offshore, walk around, meet the mostly guys that work on those platforms, those rigs, and get a sense of what they do on a daily basis."

Her job also gives her the opportunity to see the United States from a unique vantage point. "Because I travel so much, I'm looking out of airplane windows a lot. I find the scale and scope and beauty of our country breathtaking. I've been back and forth to Alaska a couple of times recently and you fly for hours and hours and hours over mountainous areas that don't have great population centers and it is incredible. It's hard to fathom the resources and vastness of our lands."

At BOEM headquarters in Washington, D.C., or "the world I live in," as Abby refers to it, she reports to "two super-strong brilliant women" – her immediate boss, Janice Schneider, the Assistant Secretary, and her "big boss," Sally Jewel, the Secretary of the Interior. The leadership team over which she presides has a number of women in prominent positions.

"When I sit and look at the table, at the small group of people I make decisions with, it's about half and half. I am very much aware and cognizant of the gender dynamics and making sure we have balance in our leadership and in the kind of people I hear from and interact with. I did that at the Maryland Energy Administration (MEA) and I continue to do it here." While she hastens to acknowledge that the oil and gas industry is male-dominated, she says she has not found that to be the case within her own organization.

Connie Gillette, deputy public affairs officer and media relations manager, says, "In fact, when we have our briefings, we have so many women scientists. I have been here for about three years and it struck me that there were so many confident and competent women in pretty powerful positions."

To encourage engineering and science majors – both male and female – to consider government service as an option, Secretary Jewell has what Abby calls "a very aggressive youth engagement plan," which includes outreach to young people from elementary school level through university. "We do science," Abby says simply. "That's really what we do here."

Connie adds, "We definitely do outreach, but it was not aimed at women in particular; it was aimed at Alaska native youth to get them more involved in their processes. Half of the youths were girls. They went to the White House and Capitol Hill. We did a lot of cool things to try to make them feel more confident about speaking up for their own community. I gave them media training and tried to help them effectively communicate what they want to convey." Perhaps in part as a reflection of the success of that program and other initiatives, BOEM was named one of the Top 50 Best STEM Workplaces for Native American professionals in the 2014 issue of *Winds of Change* magazine.

Internally, there is an active Diversity Management Program in the Department of the Interior (DOI) in which BOEM participates, but there is not a dedicated women's network within the organization.

"It might sound funny," Connie says, "but we don't have a need. It's the most equitable place I have worked."

"I have Connie," Abby says and the two burst out laughing. "I am super open to it. To be honest, I haven't taken advantage of the Women's Energy Network (WEN) or any other women's networks only because I haven't had time. I would love to, though."

Working in an organization that appreciates diversity doesn't prevent her from encountering gender bias outside of her own environment. "I see ways in which, as leader of this organization, I am treated differently, perhaps, than leaders of some of my sister organizations that are men. I still have to speak up and make sure that I'm heard. It's not uncommon that I will express something in a meeting and then a man will say the same thing and he will be heard.

"I pay attention to who people make eye contact with because you make eye contact with the person you want to influence. There are many meetings where people from the outside will meet with my deputy director, who is a man, and me, and will only look at and address him. I have been asked many times when the Director is coming to the meeting and I have to tell people, 'She's already here.'"

She cites another example. "I was somewhere; I can never remember where" – a reflection of the amount of travel her job entails – "and this man couldn't believe that I actually headed BOEM.

"He said, 'No, which *department* are you in?'

"I said, 'I'm head of it.'

"He said, 'No, which *section*?'

While Abby says she has not encountered blatant sexism in her current job or her previous job with MEA, she says it is the immediate assumptions people make about where someone fits in the world that she finds more problematic. "There are just more subtle ways that women have to pay attention to and address or they continue.

"I've been lucky working for the Governor and then working for the Secretary to have bosses that empowered me to go forth and speak on their behalf and represent their agencies, and great teams to support me. This bias is not an internal dynamic; it's more of an outward-facing dynamic."

While she has had strong support from the top down during her career, she says "the mentor question" in an interview always makes her smile. "There are people throughout my career who have been good guideposts – certainly my friend Doug Nazarian who offered me a job in the government sector that led to this amazing career that I'm having – and who have had a huge impact on my professional development, but I've never had what I perceive as a traditional mentoring role where you sit down and someone watches out for your career."

What she has had instead is a peer group of women that has come into her life at different times – high school, college, during her career in the energy industry – with whom she meets individually and "who are similarly situated and also trying to figure out this whole thing of parenting and working and being a public face and still making it home on time.

"I don't have a lot of examples that go before me to show me how to do that. Of the women here who are above me, one doesn't have kids and one's a grandma, so they don't have to figure out how to also get three kids to practice on time!"

Because she hasn't had those examples, Abby takes her responsibility seriously to show other women how to do it. And they are hungry for her advice. "There are so many women that ask to take me to lunch or to have coffee, and say, 'I want to figure out how.'"

While she doesn't do any formal mentoring – "I don't have time" – she thinks "being pretty open about the craziness of having a high-profile job and a highly demanding family, I feel like I can lead by example in that way."

Connie interjects, "And she's also doing the job very well. That's what makes her an example."

At the moment, Abby has three mentees at home – two daughters, ages 12 and 10, and a 7-year old son. "How do I do it?" she muses. "I don't know. I have a great partner on the home front and that helps a lot. I don't know that I'm particularly organized. I have a good memory; I think that's one of my greatest assets," she says laughing. "I have lots of energy; I work out almost every day."

Like everyone else these days, she is on social media and her tweets are a testament to her dedication to exercise and the role it plays in keeping her life in balance. Her tweets are often short and simple, which is not surprising given her time constraints.

"Hot yoga"

"Elliptical bound"

"Spinning"

"Running. Not in the mood but doing it anyway."

"Running to get out my frustration."

"Running before testifying."

"Running to clear my head."

"Running before another crazy day."

"I have a strong belief in balance and that guides my decisions on how I spend my time," Abby says.

When she was named by BOEM as one of five Trailblazing Women Leaving Their Mark at Interior, she modestly tweeted, "This woman leader is making a difference because of the women who walked before her."

While Abby has some sort of professional title ("lawyer") on her social media pages, most have not been updated to reflect her appointment to Director of BOEM even though she has held the position for nearly a year and a half. In her posts, she may give her work a brief mention, but her pages are full of family photos, captioned with terms of endearment for her husband and children.

Because two of her three children are girls, she is in a position to talk about STEM subjects and making them more attractive to girls from a parent's perspective.

"I think I might be a little biased. I went to an all-girls school 7th – 12th grades here in DC. I had been told by my parents, but certainly by my educational institution, that I could do whatever I wanted and that women were going to rule the world. It was expensive, but it was a great education.

"I don't have a great point of reference, but I know that both of my girls love science and engineering. My older daughter just finished her first year of middle school and her science teacher was her favorite teacher, which I was so excited about. My daughters could wax eloquent about STEM – 'Oh, yeah, they're trying to push girls in STEM.' It's very transparent what's happening, but it's great."

Even so, Abby says there is one place where she noticed a real gender difference when she visited the middle school. "My oldest daughter talked about – we used to call it "shop" – now it's called technical arts or something like that, and she said, 'Mom, this is all boys.' Even though the girls rotate through it – they rotate everyone through everything – it was clear that it was not really a place where she felt comfortable."

Her daughter liked home economics and wanted to sign up for it as one of her electives in 7th grade. "I said, 'No. I'll teach you how to cook and Grandma will teach you how to sew.' I felt like her time could be better utilized in other ways. She's taking world cultures instead."

She talks a lot to her children about what they want to be when they grow up. "We've gone from nail technician to cosmetic surgeon to farmer to lawyer to opening an orphanage in Africa to children's author, so I don't think they've quite settled on anything yet, but I absolutely feel like their options are open to them."

Abby may be exploring her own options in the coming year. Because she is a political appointee, there is a possibility when President Obama leaves office in January 2017, the incoming president could nominate a new Secretary of the Interior and, in turn, the two could appoint a new BOEM Director. Like so many others in the energy industry, she could be looking for a new job.

"I love energy policy, I love politics, and I love communications. I don't have a specific job that I can envision. I can see myself working in the energy sector on some conglomeration of policy, communications, and politics – and law. I love being a lawyer. Exactly what that means, I don't know, but it would be hard to get me out of the energy world."

♀ ♀ ♀

On January 17, 2017, Abby Ross Hopper assumed her new role as the first woman president & CEO of The Solar Energy Industries Association (SEIA).

12	**Eve Howell**
	First Woman Head of Australia's North West Shelf

"We [women] just made our own way."

L ondon. The Swinging '60s. The music revolution synonymous with The Beatles. Quite a culture shock for a 16-year old girl from the Sudan who, prior to her arrival in the UK, had never seen television.

The youngest of three children, Eve Howell was born in Khartoum to a Sudanese doctor and his British wife, who was a nurse prior to their marriage in the late 1930s. "My parents were quite remarkable people. It was a very unusual marriage – an English woman and a black Sudanese man."

Despite losing her father at the tender age of 10, she recalls, "My father was an absolute self-starter. He came from nothing, basically, and got himself educated by hook and by crook and ended up a doctor. He had great determination and there was always that attitude at home that you can do everything."

While it was very important to her parents that their children – two girls and a boy – be educated, they did not try to influence their course of study. "They never said, 'This is what we want you to do.' We were all expected to have some sort of career, but in whatever field we chose, which was the way I thought everybody was."

That perception was shaped, in part, by the strong role model Eve had in her own mother, who stayed home with the children during her marriage, but returned to work – this time as a secretary for British Overseas Airways Corporation (BOAC), the precursor of British Airways – after she was widowed.

"When my father died, she went back to work; she had to work to keep the family going. So from the age of 10 onwards I had a working mother."

Sudan's Parliament unanimously adopted a declaration of independence from the Anglo-Egyptian Condominium that became effective January 1, 1956, and Eve says, "By the early '60s, it was moving much more toward Muslim law and Arabic as the dominant language and my mother, as a working mother and a widow with three children, was finding it increasingly difficult." Having regained her British citizenship (after being stripped of it upon her marriage to a foreign national and then later having it restored by a change in the law), her mother made plans for the family to return to England.

Eve's older sister Leila, following in their father's footsteps, had trained as a doctor and already was working in the UK and her brother Ramsey, after attending Khartoum University, had received a scholarship to study in England. Eve's mother traveled with her to London, then returned to Khartoum, as Eve puts it, "To pack, pay her bills, and finalize her life in Sudan," before joining her children nine months later.

Eve, who had left school at 15 and worked at a bank in Khartoum as a shorthand typist, says, "It was a very different life suddenly."

Looking back, she realizes growing up in a multi-cultural society in Khartoum served her well. "I went to school with all nationalities – English, Sudanese, Egyptians, Armenians, Ethiopians, Lebanese, Indians, Italians and Greeks; and many religions – Islam, Christianity, Judaism, Hinduism and Paganism. We had the whole gamut, and I think it made me very comfortable with traveling and being with different types of people, so it probably did me good instead of having a very sheltered start to life."

In England, she lived with her uncle John, her mother's brother, and her aunt Bretta, another strong female role model, whom Eve explains, "Got a science degree way, way back in the '20s in the UK, which was quite unusual," and encouraged Eve – who had no idea what she wanted to pursue academically – to study science as well. In secondary school in Sudan, Eve's favorite subjects had been English literature and history, which her aunt advised her to continue enjoying as "hobbies," while focusing on science, believing it would give Eve a better chance at finding employment.

In an event that would perhaps portend the future, Eve says, "I remember my aunt telling me with enormous enthusiasm about a liquid natural gas (LNG) terminal at a place called Canvey Island in the Thames Estuary. It was the first receiving LNG terminal in the world and the English were so proud of it. They thought it was the best thing that you could dream of. Everyone was talking about how this frozen gas was coming in from Algeria, and what a difference it was making to the country, and everyone went and had a look outside the fence at the terminal."

Pointing out that it was located near a housing area, Eve says, "People now would be saying, 'We don't want this,' 'This is horrible,' 'It's going to be dangerous,' but people then were just fascinated."

"It's funny because I never dreamt then, having seen this terminal, that I'd ever have anything to do with LNG. It just went over my head really."

Reflecting on those early days of the industry, she says, "It's fascinating actually. You go back and read about it and you realize what giant steps they were in a way."

Based on her aunt's advice and still "not having much knowledge of what I wanted to do," Eve spent two years at technical college, focusing on physics, maths, and chemistry and earning her university entrance qualifications, which enabled her to pursue a degree in physics and maths. Lacking clear direction, she says, "I just thought, this is interesting and will lead to something."

She was required to take a subsidiary subject her first year and one of the options was geology, an area of study she had never been exposed to before. Intrigued, she took the course and was so captivated, she wanted to change her major to geology. The head of the science faculty – "quite an elderly man" – encouraged her to do a joint degree in geology and maths telling her, "You're not likely to get a job as a geologist."

Speculating whether or not his comment was based on the fact that she was a woman or because there was a lack of job opportunities, Eve says, "I think it was a bit of both. He was quite old-fashioned and he probably didn't think that geology was something for women," – despite the fact that three of the eight students majoring in geology were women. "Also, this was pre any North Sea or activity in England and really, if you were a geologist, you either worked for the Coal Board or the Geological Survey. Those were your options – or go overseas."

Giving him the benefit of the doubt, Eve says, "He probably did think that the career opportunities were limited."

She would go on to prove him wrong in the best way possible. "It was quite ironic him having said that I wouldn't get a job because I applied for the Geological Survey before I actually graduated and they accepted me on the condition that I'd get my degree.[50] I literally left university on the Friday and started work on the Monday."

In retrospect, she acknowledges the wisdom in his advice. "It may have been useful because I was taken into the Geological Survey to work in the geophysics section based upon the fact that I had a background in maths and some physics. It probably did stand me in good stead rather than being a pure geologist."

For the first three years or so, summers were spent in the field conducting regional geophysical mapping and surveys of the whole of the British Isles, where she recalls working with one, maybe two, other women. "We were very much in the minority. Most of the women were office-bound doing desk work." Due to harsh weather conditions, she, too, spent winters in the office writing reports detailing her work in the field.

The nature of her work would change in 1970 by which time exploration had been going on in the North Sea for several years. As Eve explains, "A lot more interest was being shown in the Continental Shelf and on the marine side, and I was involved with a few marine surveys doing shallow seismic and regional, large-scale, basin delineation type work."

The following year, the UK Department of Energy, which was monitoring the activity in the North Sea, requested help from the Geological Survey, and Eve, along with a few other geologists, was assigned to review the North Sea data. She knew nothing about petroleum exploration and was forced to learn quickly, but what she lacked in knowledge, she made up for in enthusiasm.

"It was absolutely fantastic! We were getting seismic data from all the big companies at the time – Shell and Esso and Mobil – and it was quite amazing for a young geo to actually start seeing this. I began to learn how to interpret seismic data and look at some of the logs and start to get an understanding of the essential ingredients to finding an oil or gas discovery. I found it absolutely fascinating.

---

[50] Eve graduated from King's College London, University of London, in 1967 with a Bachelor of Science (Honors), geology & mathematics.

"It was such a big deal for the UK because they were energy short. In fact, they were importing LNG at that stage. To suddenly find all this on their doorstep was just amazing."

The work itself was made all the more exciting by the rapid changes in technology. "I remember the seismic data going from very poor, single-fold data to multi-fold data where you could really start seeing what was going on rather than just stabbing around in the dark pretty much. At the same time, computers were beginning to take over and processing capability was starting to increase very rapidly."

Eve had the opportunity to spend several stints offshore on seismic vessels, including one "cruise," as she calls it, in the North Sea and the others on the west coast in the Irish Sea.

"It was interesting. It was me... and a lot of men. The very first time I went offshore, I had a few issues – not so much with the professional people – it was the seamen who were running the boat and they thought the 23-year old girl was easy pickings. I coped with them. Nothing bad came out of it in the end. I just had to be very firm with those guys. You have to learn how to conduct yourself and be strong. I guess some of the circumstances I came across from time to time could have been in any field."

Working offshore was a rare opportunity and one that would not come up again in her career, partly because to a large extent her work doing interpretation didn't require it, but she suspects it also was due to the prevailing attitude "in those years" that "as a woman, you weren't expected to be out there."

In 1973, having worked on North Sea data for several years, Eve came to the realization she would rather work more directly with the data rather than look at other people's data and, in her words, "Really not be able to do anything with it." It was then that she made the decision to leave the Geological Survey.

It was fortuitous timing as oil companies were actively seeking employees and she landed a job with Amoco, a big player in the North Sea in the early days, becoming the first professional woman in its exploration department in London.

Eve couldn't have known it at the time, but her career with Amoco would be short-lived. "That's another story – I got married and had kids," but she calls them "three absolutely amazing years.

"They provided such good training. I spent time in Tulsa, Oklahoma, at their research facility. They were really investing in young people and getting them up to speed quickly. It was such a good beginning for me.

"It was all men, but my immediate boss, Andrew Sharpe, an Englishman, and the chief geophysicist, Ed Trapp, an American, were just fantastic and they took time to help me build my confidence."

Eve remembers very early on having to make a presentation to their joint venture partners about a recommendation for a well that was to be drilled. "I had never done anything like that in my life. My boss said to me, 'You might be sitting there in front of British Gas and Amerada Hess and all these guys, but remember you know more about this prospect than they do, so just tell it the way it is and they'll listen to you.'

"During the time I was with Amoco, I was directly involved in eight wells, five of which were eventually commercial, so those were really heady days and I suppose I thought it was relatively easy. They were great years because the successes were coming so fast. We were very well paid and it was just lots of fun. We worked hard and we played hard.

"I have to say, after being in the Geological Survey and working for the government, working for an oil company was very glamorous actually."

Eve, who had met her future husband Peter at the Survey and gotten married while at Amoco, gave birth to her first daughter Kate in 1976 and, while she had six months to return to her job, she points out, "Those were the days when you didn't get maternity leave." She ultimately decided not to return to Amoco "because I wanted to spend at least that first year with my child."

Near the end of that time, Eve, already pregnant with her second daughter Jane, was offered a contract by Amoco that would enable her to work at home – an unusual arrangement at the time, something she attributes to the high demand for people and the fact that Amoco "knew me and trusted me." She was given a project to interpret some regional data from offshore Greenland and told to work on it when it was convenient even if that was the middle of the night.

"I was very, very fortunate. You don't see such flexibility even now, although you might see flexible hours." Eve believes giving mothers more flexibility where their children are concerned might help attract more women to the industry because she believes having to choose between family – particularly when children are pre school-age – and career "does sort of tear you up."

Her next contract came in 1977 from Exploration Consultants Ltd. (ECL), a UK consulting group to the oil and gas industry. Again, there was an abundance of work and not enough qualified people to do it, so the company agreed to let her continue working from home only asking that she spend half a day a week in the office meeting with her colleagues and providing a progress report on her work.

For the next four years, Eve would interpret data, not just from the North Sea and the Irish Sea, but from all over the world. "I did a lot of work on some of the early offshore China data, Jamaica, Vietnam. I didn't actually go to these countries – I was in the UK – but I looked at their data. I did take the family for a month and go on a little project in Barbados, which was fantastic.

"When we moved to Australia in 1981, my children were five and three and I'd had those years, obviously working hard, but being around for them. Being home and being able to take them to kindergarten and pick them up was good."

It helped having a husband who didn't mind reversing roles. The industry was booming and exploration in Australia was beginning to gain momentum, already having seen success offshore from the Woodside discoveries in the '70s. When ECL asked if she would go to Australia for three months to help open an office, her husband was so enthusiastic about the prospect, he offered to quit his job at the Geological Survey and stay home with the children for a few months. The family then embarked on what Eve calls "a big adventure."

While assisting in the opening of the ECL office, she simultaneously had a three-month contract with Mesa Petroleum working as an interpreter. The contract would eventually extend to 18 months and her work with ECL would last for three years, by which time the family had fallen in love with Perth, and Eve had begun running the ECL office.

"That was really my first start being a manager. I found it quite difficult because suddenly I had to be accountable for the revenues, and making sure we got paid, and making sure the head office was happy that we were actually making money, but you have to learn."

However, she found consulting to be disruptive to family life. Because the company was doing work in Asia, she spent a lot of time traveling to Jakarta and Singapore, something she was not comfortable doing while her children were still so young. Eve found herself longing for the stability of a corporate job and thought she had found it in 1984 when she joined Occidental Petroleum in Perth.

"Occidental was a company that I'd admired in the North Sea because of its early days with the Piper and Claymore fields. They had just made a couple of discoveries in Australia and were seen as real sort of go-ahead people. They promised me the world."

Six months later, the job was over.

"They had their big discovery – Caño Limón in Columbia – and decided they were going to put all their efforts into that. They put their Australian assets up for sale. I thought I might have to go back to the UK, but it didn't happen that way. Their assets off the North West Shelf of Australia were sold to Bond Corporation. Alan Bond, the Australian entrepreneur, who won the America's Cup the year before, had a petroleum division and he was very keen to grow it, so he took most of the people across." Eve suddenly found herself the chief geophysicist at Bond Petroleum.

"We actually developed Western Australia's – not Australia's – first offshore oilfield, which came on in '86. It was a relatively small field – 50 million barrels[51] – called Harriet in the Carnarvon Basin, but that was quite a big deal to be the first one to be producing an offshore oilfield in Western Australia."

Production started during a downturn in the industry, which Eve recalls as "a miserable time" when oil was less than $10 a barrel and large numbers of people were being made redundant. One of the keys to her survival she believes is that she had a broader range of experience than just as a pure geologist. "There was quite a shortage at that time of seismic interpreters and that was probably in my favor." She would see other downturns during her career and says, "I probably was shielded a bit more by the fact that I was working for companies where I had a pretty deep knowledge of all their assets."

In 1987, the same year the stock market would crash, she was promoted to exploration manager. She would stay until 1989 when there was a corporate collapse – petroleum was only a small part of the Bond empire – and the company went bankrupt.

The Australian assets were put up for sale and eventually bought by a consortium led by a small company out of Oklahoma called Hadson [Energy], which sent people to Perth to run the assets. Eve was retained as the exploration manager due to her familiarity with the assets and remained with Hadson for four years until it was acquired by Apache Corp. in 1993.

"So there I was again. The company disappeared on me! In rather quick succession, I went from Occidental to Bond to Hadson to Apache basically with pretty much the same assets," which was to her advantage

---

[51] A press release by Carnarvon Petroleum Ltd. dated May 24, 2014 stated, "Cumulative production from the Harriet fields has exceeded 100 million barrels of oil and condensate." (Eve's comment refers to an earlier period of time.)

as she knew them so well. After all the upheaval, she appreciated being with a well-funded company that had the resources to pursue an active drilling program, although by 1995 she was questioning whether she wanted to remain an exploration manager for the remainder of her career.

Telling her immediate boss she believed it was time for someone with fresh eyes to take over the role of exploration manager, Eve proposed transitioning into business development. "I said, 'I'll do my bit as well; I'll do an MBA.' It convinced Apache that I was serious about wanting to change direction and I was prepared to give my time to it."

The company, obviously eager to retain a valuable employee, funded the program, which she did part-time by distance learning through Heriot Watt University, while continuing to take on more responsibilities in business development.

She doesn't necessarily believe that having an MBA is a "prerequisite" for women getting ahead in the industry, saying, "When I did my university degree, I didn't do anything but technical education. There was nothing that taught me about being in the business world. It was just geology and that was it."

What she does believe is that it gave her *credibility*.

"I remember one of our joint venture partners, a finance guy, was very scathing when Apache moved me into that role and said, 'What would a geologist know about business development?'

"If you've got the letters, you can say, 'Well actually, I have a business degree, so don't try that on me." She laughs, adding, "This wasn't one of my favorite partners."

Proving it is never too late to add "a few more strings to the bow," Eve says, "For me, it was a real benefit doing my MBA late, over the age of 50. A lot of people will do their MBAs very early and very young and that's fine if you're going straight into business. But if I'd done it when I was young and then gone back to being a geologist or a geophysicist, I think it would have just been a waste of time really."

She gradually moved into business development and found herself getting involved in oil and gas marketing, as well as the overall business strategy. Farm-in and farm-out deals and acquisitions also fell under her purview in business development.

Eve was made deputy managing director in late 1999, which she calls "a good step," and by 2004 she had been promoted to managing director and regional vice president. Prior to that, Apache had always had expat managing directors and regional vice presidents in Australia, so calls it "a moment of pride" to have been named the first non-expat. And the first woman. And to report to the head office, something she had never done before.

Two years later, in 2006, "I had just passed 60, so I thought it was probably time to hang up my boots," she says laughing. "If you take all the years I had been working on the same assets, it was a long stretch. I was getting a little bit itchy to do something else... even if that something was retirement."

The one thing she never had in her long and storied career is a female mentor and that she says, "Is a pity. There wasn't anyone in the sort of position that could have helped me. We [women] just made our own way."

Perhaps because of that, Eve is willing to share the knowledge and expertise she gained during the course of her career, which spanned over 40 years. "I think mentors can be important. I've got a couple of younger women at the moment, who ring me up every now and then and say, 'Can we have a cup of coffee? I'd like you to help me with this situation.' These are both women with a lot of potential, and it would be wonderful if they would stay in the industry and develop that potential.

"I find it quite disappointing that there aren't more women in the industry and I'm not seeing many women at all in senior positions."

After leaving Apache, "I had no intention of going back to full-time work at that point. I thought I might do some consulting, I'd try and get some director roles, do something a little bit different." Eve has gone on to hold non-executive director positions with Buru Energy Ltd., Mermaid Marine Australia Pty. Ltd., Downer EDI, and others.

When she was approached by CEO Don Voelte to go to work for Woodside, she was ambivalent. That is, until he told her he needed somebody to be in charge of the North West Shelf.

"The North West Shelf was such an iconic project in Australia. It was – and still is – Australia's biggest resource project."[52] After thinking about it for a while, Eve was still undecided, but ultimately concluded, "This is fantastic; I can't turn this down."

She agreed to work for Woodside for three years "and I stayed for five and a half, by the way."

Throughout her career, the projects she had been involved with had been conventional oil and gas; she had never had any exposure to LNG (aside from marveling at the terminal on Canvey Island as a teenager).

---

[52] Source: Woodside

"All my life, I'd just liked doing something a bit different, a bit new. The more I thought about it, I realized, 'This is actually an extraordinary opportunity.'

"It was such a complex project, such a huge operation, but the thing that was really interesting for me was all the commercial connections because we were exporting LNG to Japan, to South Korea, to China, and developing all those relationships. Within one week of joining, I had to go to Japan. I'd never been to Japan and I made countless trips there with Woodside. I went to South Korea quite a few times and to China."

Despite the perceived glamour of international travel, the first trip to Japan was an inglorious start to her new position.

"I was kind of dropped into the role because we'd had a real problem at the plant with one of the compressors, and I was the poor guy – poor person – who had to go and apologize for the fact that we'd missed our shipments. It was not my fault; I was the new guy." Fortunately, she can laugh about it now.

Because it was such a high-profile project, Eve says interactions with the Australian government were very strong. "Both the state government and the federal government in Canberra took a lot of interest in this project. To go to Canberra and get welcomed in and meet the prime minister and talk about your project was incredible. I got to meet four of Australia's prime ministers... because we keep changing them!"

The company went through a major expansion on the operational side, adding a fifth train to the LNG plant, building new platforms, in addition to developing gas resources. And, as Eve recalls, billions of dollars' worth of new investments were going into the project as well. However, she found the most challenging and interesting aspect to be the joint venture itself.

"It was quite a complicated JV and Woodside, as an operator, had only 16 per cent of the project and we had BP, Shell, BHP, Chevron, Mitsubishi, Mitsui, and the China National Offshore Oil Corporation (CNOOC) as partners. The dynamics were just amazing. It was challenging, but we made it.

"I felt very privileged to have been able to do that. I was also very proud that I was the first head of the North West Shelf who wasn't a Shell person. And I think I can also say who wasn't an engineer as well... and the first woman, of course.

"It was an extraordinary end to what's been a wonderful career."

♀ ♀ ♀

In September 2018, Eve Howell was inducted into the Offshore Energy Center Champions Hall of Fame in Houston, Texas.

	Dr. Amy Jadesimi
13	**Managing Director,**
	Lagos Deep Offshore Logistics Base (LADOL)

"It's going to be a while before you get the moderating influence of diversity changing the offshore culture."

She grew up in London more than 3,000 miles from her birthplace of Lagos, Nigeria, and yet Amy Jadesimi says she identifies strongly with her Nigerian heritage. Having left the West African nation at the age of four, she only has what she calls "childhood memories," but says her family has a strong heritage that enabled her to stay connected culturally. Growing up, she was aware of her Nigerian ancestry, which includes a relative who was the biggest palm oil trader in the world when palm oil was like crude is today; her maternal grandfather who was Nigeria's first finance minister; her paternal grandfather who was an Anglican bishop; her mother who was the first female Nigerian journalist for Radio Sierra Leone in the 1970s; and her father who set up the Arthur Anderson accounting firm in Nigeria.

"Coming from a family like that, it was natural that at some point I would want to go back and see what I could do, if anything." However, it would be in her "next life," as she is fond of saying.

Educated at British boarding schools, Amy refers to herself as a "sporty geek" who played netball and tennis and liked the sciences, but her favorite subject was maths and she found particular enjoyment "laboriously working through problems." Although she wasn't sure what she wanted to do professionally, she says it only seems natural given her interests that she would go to medical school. She chose to attend her father's alma mater, Oxford, where he had earned a Master of Arts in jurisprudence.

"I have to admit, I didn't have a burning passion to be a doctor, but I liked studying the sciences and, at that point, I didn't have a burning passion to do anything else either. Once I went to medical school, I found it fascinating. I wouldn't have stayed and gotten a doctorate if it wasn't something I grew to have a passion for." Incredibly, she did an internship at Island Maternity Hospital where she was born in Lagos.

She had an epiphany her last year of med school while working in a British hospital. In talking to the senior physicians, she realized that most doctors, particularly in England, make the decision to go into medicine at the age of 15 or 16 and stay with the profession for the rest of their lives.

As someone who has always been interested in many different things, when she graduated from Oxford and received a job offer from Goldman Sachs, Amy thought she would work there for a year or so and then go back to medicine "just so I would have done something else in my life. I thought I might as well have as many experiences as possible." Medicine, she says, is forgiving in that way; you can leave it and return later. Logically, it would be easier to do in the beginning rather than halfway through her career as a doctor.

She joined Goldman as a financial analyst and, to her surprise, she loved it. Working in mergers and acquisitions in corporate finance, she dealt with transactions in pharmacology, health care, and technology, as well as general industry deals, giving her broad exposure to the financial world and an interesting view of life.

A self-described workaholic at heart – "I still haven't decided if that's a good thing or bad thing" – Amy says she didn't mind working *a lot* and being in the office behind a desk. As someone who, by her own admission, is not always comfortable in social situations, she also appreciated the camaraderie that developed in the cloistered environment. "You have your Goldman family and then you have everybody else, who you hardly ever see because you're always with your Goldman family," she says, laughing.

In her third year, Amy was promoted to associate and faced another tough decision – whether or not to go to business school and, if so, which one. She seriously considered going to Harvard but determined if she was going to take two years out of her life to attend business school, she wanted it to be a "really different learning experience." Harvard seemed similar to Oxford and she felt it was more driven by academic achievement; whereas Stanford was more driven by personal achievement, "But not in a trivial way. I felt like people weren't trying to achieve external goals anymore. They were comfortable that they didn't have anything to prove to themselves or other people in that way."

She calls it a place where there was "unpretentious learning. It was really just about figuring out who you wanted to be and very seriously and studiously honing your craft. There are a lot of classes at Stanford that are really about interpersonal dynamics and about learning how to lead in terms of emotion. There are a lot of soft skills you learn there, which I'd never studied before, having been very academic and focused on the sciences, medicine, and then finance work at Goldman. Stanford was definitely the school for me just because I wanted a different academic experience."

Amy describes herself as "really torn" even after being accepted into Stanford, and two weeks before she was due to leave Goldman – "they keep you working right up until the end" – still wasn't fully committed to the idea. "I was half-thinking, maybe I should stay here and work my way up." One of the partners at Goldman convinced her she could always go back and that it would be more advantageous to earn her MBA, learn a broader range academically, and have other experiences for a couple of years before returning to Goldman.

She equates it to leaving medicine to go into banking. "It was the same sort of thinking – 'Let me have this experience and it will be a positive experience in my life' – and that's why I wanted to go to Stanford as well."

Prior to graduating from Stanford, she typically only spent about 10 days a year in Nigeria during the Christmas holidays visiting her maternal grandmother, who no longer travelled much. After completing her MBA, Amy felt it was time to go back to Nigeria and "really give it a go. I felt I had sufficient experience and learning that I could contribute something. I grew up thinking about Nigeria and understanding the situation there, so although I wasn't 100 percent sure what I was going to do, I felt like I was ready to move back and achieve something."

Having said that, she did not go back with the express intention of working for her father's company, LADOL.[53] Conceived in 2001 as the

---

[53] Oladipo (Ladi) Jadesimi is executive chairman of LADOL

first logistics and engineering base west of Lagos for deepwater offshore oil and gas projects,[54] the development of LADOL had just started when Amy arrived in Lagos in 2004. At that stage, her father was putting together the company's first formal business plan and she casually offered to take a look. "Having worked at Goldman, I had some skill sets I could help them with."

Two weeks later, she found herself in South Africa where management was talking to a major petroleum company about investing in LADOL. She spent the next few weeks in Cape Town building the business model. From that point onward, she became more and more involved with LADOL. "Obviously, I had heard about it and I understood what my father was trying to achieve, but I think at that stage neither of us realized the extent it would take over our lives for the next decade," she says, laughing.

The development and expansion of the company haven't followed a linear path and Amy says there have been many ups and downs over the years. "In some ways, what we do is very pedestrian; we're building infrastructure. The infrastructure adds up to an industrial town, which is also a free zone."[55] She notes that each phase of the development of LADOL has almost been like running a different business, which means her job as managing director is constantly changing.

"When I started, it was almost like I was doing a financial consultancy, putting together a business plan, an information memorandum, to raise some money. Then we went into a construction phase and I was part of the team that was looking at tendering, selecting contractors, putting contracts in place. Developing the base was the building phase. Once it was built, we then became operational and we were running a logistics service provider and a rig/vessel repair company that was also a provider of hotel services and office services.

"As we continue to develop the free zone, it is always necessary for the company to raise funds. While the original business of logistics and vessel repair remains one of its core focuses, the company has expanded into shipbuilding. The first ship is being built now with Samsung Heavy Industries, which will partially build and integrate a (Floating Production Storage and Offloading) FPSO in LADOL, which is a huge

---

[54] Source: LADOL

[55] An area within which goods may be landed, handled, manufactured or reconfigured, and re-exported without the intervention of customs authorities. Source: Encyclopedia Britannica

achievement for the country. Nigeria has been trying to integrate an FPSO onshore in Nigeria for decades."

Although running a large, diverse company is challenging, Amy admits it has fulfilled a professional longing. "I loved being a banker, so I really enjoyed my years at Goldman – and I know that's not a very PC thing to say these days – I did like the finance environment, but one of the things that always struck me as 'slightly odd,' I think is the best way to characterize it, about our job is that I didn't feel like we were necessarily building anything. I wouldn't go as far as to say we weren't adding value – in investment banking, you decide the fate of small and sometimes very large companies – but you're not involved in the personnel, the production, what that company actually does, and so there was always something in me that found that a little bit strange."

In working at LADOL, Amy seems to have found what was eluding her. "All we do is build. What we're doing is very tangible. We're affecting people's lives. It's a very hands-on company. Even though we're getting bigger, you can see the impact every day of what you're doing. One of the great things about working in a fast-growing market like Nigeria is, the opportunities are plentiful and, yes, it's tough, but having worked at Goldman, I can tell you that working in London in the city is also very tough! In a way, anything you do, if you're going to do it well, is going to be hard work. Working in a growth market, you can set up a company that changes the way people do business in a region within a generation." She compares what LADOL is doing "in a small way" in the oil and gas sector to what MTN[56] has done for mobile telecommunications.

While she is appreciative of the publicity LADOL receives – "the size of the project, the amount of money we're investing and what we're building, for sure that's unique" – she feels there are other stories just as deserving.

"What I've found, moving back to Nigeria, is that my faith in humanity has really been restored. Working in Lagos you see people who work multiple jobs, you see people who, at different points in time, are surrounded by a semi-chaotic state, but for some reason things work. It's as if people have a very strong natural desire to make the best of any situation and bring order out of chaos. I find Nigerians have enormous

---

[56] MTN Nigeria is part of the MTN Group, Africa's leading cellular telecommunications company. Source: MTN

patience, the best temperament. Maybe if we had less patience, we would have demanded democracy sooner," Amy says dryly.

She sees LADOL as completely disrupting the market – "and oil and gas is a very difficult market to disrupt. There are a lot of very powerful, wealthy, corrupt people who basically have gone out of their way trying to keep us out of the market because our presence adds transparency and value, making it harder for them to continue to profit through rent seeking. So, for us to have survived and to be where we are now means that Nigerians are inherently good and that the system works – eventually.

"There are people who have risked their jobs – there are people who have lost their jobs – there are people who have had their lives threatened but have still helped us because it was the right thing to do. Literally, everything – all the external incentives – was telling them they should not help us or give us a certain approval even though we were entitled to it, and in the end, they did because it was really the right thing to do.

"Just experiencing that, of course, touches you and makes you more passionate about what you're doing because it's so real. There were times we weren't sure LADOL was going make it. And in those times, I thought, 'I'm so glad I've done this because I've had these amazing human experiences.' Apart from what we've achieved professionally and what we've achieved as a company, it's really touched me. Definitely it's changed my life and my outlook. Like I said, it's really restored my faith in humanity."

Amy realizes, however, that is not how most people perceive Nigeria. She is not one to shy away from acknowledging the negative perceptions and doesn't think anyone else should either. She equates it to the way people looked at China 30 years ago or the way they viewed Norway, which may seem like a surprising example, but she says, "People forget now, but in the 1960s, and early '70s, when Norway first started exploiting its oil and gas reserves, it was the British that were going over there, showing them how it's done, and the British thought of them as kind of a bunch of backward fishermen. So every country, every culture, goes through phases where, for whatever reason, it's not positively perceived by the outside world. What the countries that have changed that perception have in common is that they have done it themselves." While Amy doesn't think it is necessarily reasonable for people to perceive Nigeria the way they do, she feels there are a lot of things Nigerians still have to do to change that perception and show the outside world that the reality is very different.

She believes the solution lies partly with controlling their own media and partly with what the government does but also says a large part of it has to do with what the private sector does, particularly in terms of investment, and how Nigerians lead by example in their own society.

Amy says it is coming back into fashion in Nigeria to look back at one's ancestry and understand that the assumption that the way some members of the political class, as well as some prominent businessmen, behave today is the norm is absolutely false.

"Nigerians who built Nigeria were honest and patriotic men and women, who were vehemently opposed to corruption. In fact, a hundred years ago the worst thing you could accuse a Nigerian of was stealing, and that was the norm for most of our history. We need to learn our real history and teach it to the generations coming up now."

She cites the example of her ancestor Chief Nana Olomu (1852–1916) one of Itsekiri's greatest leaders and heroes, who used his intelligence, ingenuity, and bravery to gain control of the palm oil trade, a competitive and complex industry. He was also a state builder, having built a new capital at Ebrohimi in the Delta region, which he sand filled and developed from a Greenfield swamp into an impregnable capital city.

As Amy explains, "Foreign commercial interests were threatened by Nana's grip on the palm oil trade and his unwillingness to accept one-sided trading terms. Europeans laid siege on his capital of Ebrohimi, but Nana was a brilliant military leader who successfully repelled the European armies on three occasions when they were forced to withdraw with 'heavy' casualties. Nana's capital eventually fell on September 25, 1894, largely due to Dogho, Nana's local rival, providing the Europeans with logistic and intelligence support, including showing them the best route to Ebrohimi.

"The poignant thing about that story," Amy says, "is that Nana was defeated by divide and conquer tactics. So they didn't actually defeat him militarily. They defeated him by dividing his people. Ultimately, we defeated ourselves, and in some ways that continues today." Although her ancestor was expelled from Nigeria, Amy is still impressed by what he achieved as well as the society that existed at that time.

She has come to realize the importance of telling those stories and celebrating them, so that the young people growing up in Nigeria today – and even the older people – have those stories to relate to and counteract how the media – including the local media – "defines our society today, which is in largely negative terms. The same thing can be seen to an extent in Western culture. Certain things, such as reality TV, which perhaps don't show the best side of humanity, don't have as corrosive effect on Western culture as they do on ours because the history is so well known and so well taught in schools. We don't know our history well enough to defend against the negative stereotypes in popular culture.

"I'm very proud of what we've done at LADOL so far, but I definitely hope it's one of many such success stories across Nigeria. Part of our vision is that our success helps to shine a light on and encourage support for similar developments in other sectors and/or on a smaller scale. There are many companies operating transparently and honestly that have been investing in Nigeria for a very long time, and they have found a way to overcome the challenges you encounter in a growth market and be successful. Supporting long-term, hardworking investors is where the real opportunity lies in this market."

Amy has said she wants LADOL to be replicated – and well diversified – across Nigeria by 2025. And those challenges she mentioned? "I hope by 2025 we won't have to deal with rent seeking kleptocrats trying to keep the market small and inefficient."

She doesn't believe any business can be truly successful if it doesn't have diversity in its staff and management, and as a financial analyst she knows the numbers back that up. Diversity has an economic impact on individuals, businesses, and society as a whole.

"From a business perspective, diversity throughout the whole free zone is really important. I don't think there are any areas that I would exclude women from. If you look at what we're building now, you could think of LADOL as an infrastructure development company a little bit like Dubai Free Zone, where we're building out the road network, underground reticulation, sewage system, etc. You can also think of us as a utilities company where we're building this 24-megawatt power plant. You can think of us as a deep offshore logistics support and ship repair company, fabrication, and ship building company. Within all those different activities, there isn't anywhere women can't contribute. A woman can captain one of our supply vessels, a woman can run our shipyard business." The challenge, Amy says, is finding and training those women.

She believes that is where a type of "positive discrimination," comes in. It doesn't mean that a company is not going to hire qualified men; rather, if a company accepts the fact that there is an economic benefit to diversity, it has to invest extra money in trying to achieve that diversity.

She says the "positive discrimination" comes in the sense of going out of the way to try to find qualified women or to train women who could move into those positions, something LADOL is already doing.

"I never really understand why there's controversy around this. I always try to take the emotions and politics out of it and I just look at it as a business. This is private sector. If I know that having a workforce that's highly diversified is going to make me more money, I'm going to invest in technology and training programs that allow me to do that."

She says Nigeria faces the added problem that on average women are more excluded from primary and secondary school and later university education. "If you look at the socio-economic breakdown of Nigeria, women are still grossly, grossly disenfranchised. Women aren't as educated as men because they don't have access, which makes the job of employing women even harder." By the time a company is hiring graduates, the pool of women it has to choose from is smaller than it would be internationally.

As the private sector demand for women increases, Amy believes there will be more pressure on government and families to educate women. "I'm hoping that we will see the impact of that in the next ten years. But it has to start with growing the 'real' private sector and by real private sector I mean companies that are adding value, that are investing, that have good corporate governance, that are making their money by producing goods or services in Nigeria, as opposed to collecting agency fees, fighting to maintain the status quo, and not adding to the GDP." She believes growing the real private sector is a very effective tool in pushing government to do more in terms of educating women as well as helping to empower women.

The main thing holding women back is the lack of opportunity, she says. "If you look at the history of the world, there's never been a point at which men freely opened the doors and welcomed women in. Even though there is an economic rationale for diversity, it took 100 years for the financial sector to recognise that. I am sure thousands of women had to fight their way into the financial sector before they could even do the analysis that has now revealed the huge benefits of gender equality in the workplace. One study showed that having equal representation of both sexes in the workplace increased top line revenues by 41 percent.

"The financial sector is always more innovative than general industry, so I expect it will still take years to reach a 'critical mass' of qualified, trained women in certain jobs. Particularly, in the case of offshore work, if you're talking about sexual equality in engineers on drilling rigs, well that could take another decade."

While Amy sees more women attending university and going into the field of engineering, many of them are being educated outside of Nigeria. She says part of the hope is that those women can be attracted back to Nigeria. "The other side of the hope is the real private sector gets so big that there's such a huge demand for these women that more emphasis is put on educating women at all levels in Nigeria."

While it is a daunting task and many things have to happen for it to become reality, she believes her country is headed in the right direction. "From a private sector perspective, I'm 100 percent certain it's just a matter of time because the economic rationale is there, and sooner or later businesses do what is economically rational. We just want it to be sooner, so we have to look for ways to make equality happen in the next ten years."

It is something she is forced to deal with even within LADOL. The company's base is an island free zone inside Lagos Harbour and the men who work on the vessels live in LADOL, creating an almost exclusively male-dominated environment.

"On the one hand, I think there is a historical offshore culture, which means that we had to do certain things which I didn't expect," Amy says. "For example, we don't use any female junior staff. The free zone has a hotel and recreational areas, where clients relax and sometimes stay for several weeks at a time. The only females we let live in the free zone are a little bit older and able to handle themselves. I think it's going to take a really long time for that aspect of the job to change because this is how people act in their leisure time; it's really what the offshore culture breeds, and again we need the number of women to reach a critical mass to change that culture.

"Typically, women get separated into two categories. I guess the category I get put in, along with women that live and work on the base, is that you're 'not really female.' That's kind of how I look at it. I don't think they necessarily register me as a female because they don't interact with me and the other senior women the same way as they do other women. That's a very crude way of looking at it and a horrible generalization. Obviously, there are some great guys. I've never felt uncomfortable in that way. However, it is literally 99 percent men, so it's going to be a while before you get the moderating influence of diversity changing the offshore culture."

Amy says the amount of media attention paid to prominent women in the petroleum sector in Nigeria and Africa as a whole makes it seem like there are greater numbers than there actually are. "Statistically, it's about one percent. It's very rare to go to a meeting and have another senior woman in the room. Perhaps in finance it's different but in oil and gas it's definitely extremely, *extremely* male-dominated."

She reiterates the importance of Nigerians knowing their history and giving people – particularly girls and women – positive role models they can admire and relate to. "It's the same with women all over the world. It touches me when girls tell me about having those role models and understanding the path those women followed and how they became successful." Without having accomplished women to look to, she says it is difficult for women to know how to overcome the challenging situations they will encounter in their lives and careers.

Amy cites her own mother, Alero Jadesimi, not only as a model of professional success but personal strength and dignity. Alero's father, Chief Festus Samuel Okotie-Eboh, Nigeria's first finance minister, was assassinated in a military coup in 1966 when she was 21 years old.

"She lost her father in this horrific way when she was young. She went from a life of extreme privilege and being very protected to a very tough period where she achieved on her own and carved her own path by becoming a journalist and then a broadcaster; in fact, she was the first and only Nigerian female journalist working for Radio Sierra Leone in 1971.

"Apart from her professional achievements and her intelligence – she is one of the smartest people I know and a constant source of advice and information – she's also a very compassionate, kind, and beautiful person. So growing up, I had that constant example of someone who had been through a horrific experience, been abused and maltreated, but who rose above all that and remained a compassionate and generous person. I don't know if I can live up to that example, but having it in my life has been inspirational."

After a successful career in broadcasting, her mother went on to run a charity that was part of the Prince's Trust. Its charitable model was a form of micro-finance to enable women to start their own businesses.

Already on her third career, Amy envisions doing other things after LADOL. Now 41, she says she definitely would like to have a family at some point. "I guess I'm not one of these people that worries about things outside of my control, so I don't know when that would be exactly."

In the meantime, she has other professional goals. "At some point I would like to take some time out to do something completely different, something that's not finance and infrastructure-related. My next life might be a more academic one. I think it would be interesting to learn about other languages, other cultures. I'm very fascinated by history because I think it can teach us a lot about today. We've forgotten a lot of our history and that may be why we keep making the same mistakes. There are lots of things I'm interested in outside of this very commercial, serious world I've gotten myself into. And of course in the back of my mind, I sometimes flirt with the idea of going back to medicine in some way, shape or form."

In reality, she already has. While in the US, she met Martha Madison Campbell, PhD, the founder and president of Venture Strategies, an international non-profit focused on health and development. When Amy moved back to Nigeria, Venture Strategies wanted help launching a maternal health initiative, which she thought was "rather brilliant" because it was about using the private sector to achieve a positive social outcome. Who better than an Oxford-trained physician to become its Nigerian representative?

As Amy explains, women were dying due to postpartum hemorrhaging (PPH) after childbirth and needed access to the drug misoprostol to help control the bleeding. The commonly used treatment at the time was oxytocin. Oxytocin is not a suitable solution in Nigeria because it has to be refrigerated and injected. Misoprostol is as effective and can be taken like aspirin. Nigerian gynecologists actually recognized the need for the drug and reached out to Venture Strategies to help them register and distribute it. Thanks to Venture Strategies and Amy, Nigeria became the first market in Africa to register misoprostol for the treatment of PPH. The drug was distributed locally by Emzor Pharmaceutical Industries, Ltd., the largest indigenous pharmaceutical company in Nigeria. ("As it happens, Emzor is run by a woman" – Stella Chinyelu Okoli (OON[57]) founder and CEO.) Stella Okoli and her company supported the misoprostol initiative and helped it become so successful that her company now manufacturers the drug locally, putting it in the hands of thousands of women who would otherwise have no treatment at all.

"It's been a complete success," Amy says. "The aim was to get the private market to achieve a sustainable development goal, i.e. reducing

---

[57] Officer of the Order of the Niger

PPH and thereby putting in place a sustainable solution. Once we achieved the goal of having the private market take over and even manufacture the drug locally, that was that. But I would love to work on another similar project." While she is no longer involved with Venture Strategies, she says, "Medicine is definitely something I see myself doing in my next life."

Wherever this life takes her, Amy carries with her the words of Archbishop Desmond Tutu. In 2013, she was chosen to participate in Tutu's Children, a five-month leadership program conducted in South Africa and London, and filmed as a four-part television series by Al Jazeera, in which participants promised to dedicate themselves to transforming the African Continent.

Amy, who describes herself as "not an emotional person at all," breaks down in tears when telling the Archbishop she remembers hearing him speak at Oxford in 1994 during her days as a medical student. "I just wanted to thank you because it really was the beginning of me regaining my pride as a black person."

Archbishop Tutu tells Amy and the group, "It is believed that you are people who can make a difference" (Al-Jazeera, 2013).

14	In Memoriam
	Zara Ibrahim Khatib, PhD (1953 - 2014)

**"We have a verse in The Quran which translates:
'Do not hate a situation, you never know it may be for your best.'[58]
This is what I always hang [onto] and [it] always turns to the better."**

These are the words Zara Khatib[59] wrote in an e-mail to one of her Shell colleagues, who was also a personal friend. According to her friend, this verse was one of Zara's personal mottos.

Born in Lebanon, Zara earned both a bachelor's and a master's degree in chemistry from the American University of Beirut. She also held a Master in Biochemical Engineering and a PhD in Chemical Engineering, both from the University of Wales. She later became a US citizen.

She served on a United Nations Sigma Xi committee on sustainable development working with a distinguished team that included several Nobel laureates. She also presented to the United Nations on developing the involvement of girls and women in industry and technology. A two-time SPE Distinguished Lecturer, she was also a member of the G8,[60] and served on the World Energy Council on Clean Fossil Fuels.

Zara began a long and illustrious career with Shell Oil in 1984, which culminated in her role as chief technologist at Shell's Gas Research Centre in Abu Dhabi, United Arab Emirates. She retired from Shell after

---

[58] The Quran, Surat Al-Baqarah 2:216

[59] Zara Ibrahim Khatib: born Jan. 25, 1953, died Sept.30, 2014).

[60] Group of eight highly industrialized nations—France, Germany, Italy, the United Kingdom, Japan, the United States, Canada, and Russia—that hold an annual meeting to foster consensus on global issues like economic growth and crisis management, global security, energy, and terrorism. Source: Council on Foreign Relations.

Photo Credit: Paul van den Hoek

nearly 30 years. At the time of her passing, Zara was managing director and partner in Smart Energy Engineering.

Here, she is remembered in the words of her family and friends.

Corbett Legg, Zara Khatib's husband:

I think that what made Zara unique was finding value and capabilities in many people that had not previously had the opportunities or encouragement to progress their careers before their involvement with her. This is what she was the most proud of, and of course Shell benefited from this increase of value of the personnel as well. And the beneficiaries of her attention and encouragement were not just women. Although she spoke to probably every women's group that asked, she never considered herself a women's rights champion. If she saw capabilities that would be useful on a technical level, she completely ignored any present or previous history or perception that other managers might have. She considered that recognizing potentials, believing in someone, providing the opportunities, encouraging their success, and insuring recognition of their contribution as being the most important thing she could provide. As a Muslim, it was really a religious obligation to her. She spoke to and encouraged many students and young professionals, both male and female. It is just that, especially in the Middle East, this encouragement to female engineers and scientists by a woman that started out just like them and had risen to a level of respect within the worldwide technical community was new. Every young female Arab technical professional that I have met that had heard her speak can still relate the positive affect that brief experience had on their direction and subsequent success.

Paul van den Hoek, Shell colleague, Water to Value team member, and personal friend:

Zara was the "mother" of the team. Those to whom I spoke after the sad news of Zara's passing away all mentioned to me that their time with Zara in the W2V team was the best time they ever had in Shell! She really cared about her staff. For example, by walking around the work floor very regularly and discussing with people what they were doing and why. Also, by thinking about their careers and helping them with getting a promotion (e.g. myself; but there were quite a few others as well). At times, she could be very direct and critical towards her team members (earning her the teasing nickname "Maggie").[61] I remember certain instants where we were involved in shouting matches over the corridor. But... the nice thing with Zara was always that she wanted to make things good afterwards, and so after each little quarrel we generally were the closest of friends again in no time! At one moment, she started to make me from a 'technical worker' into a 'team lead' and very quickly [made] a couple of very helpful suggestions of what I could do as a 'good' team lead.

In a way, you could argue that Zara's strength was at the same time her weakness. She cared for the team and the people within it, and she stood for her team towards the outside world. That is of course quite contrary to what one sees nowadays with many leaders.

Paul Verbeek, Shell colleague, Water to Value team member, and personal friend.

In May 2015, Paul Verbeek co-authored a presentation, which he gave at the Society of Petroleum Engineers (SPE) Produced Water Handling and Management Symposium in Galveston, Texas. The presentation Turning Water to Value: The Embodiment of a Vision was a tribute to Zara Khatib.

In his introduction, Paul says, "Her knowledge, creativity, passion and interpersonal skills have created value in business and in people and for the environment. 2001 – 2004 Water to Value team leader (Shell) integration of subsurface and surface knowledge was the driver behind Zara's water management initiative."

---

[61] Reference to British Prime Minister Margaret Thatcher.

And in his closing remarks, he states, "It takes the vision, passion, and creativity of people to deploy fit-for-purpose solutions. There is no silver bullet to fight water.

"Thanks to people like Zara Khatib, the industry continues to progress."

Jip van Eijden, Shell colleague, Water to Value team member, and personal friend:

Zara Khatib was the mother of Water to Value. I have a vast resource of memories of the years that I worked for Zara. The strongest memories, however, go back to all the exhausting but always very inspiring and enjoyable business trips we did. Exhausting because keeping up with Zara's energy and drive was not easy. Enjoyable because it was always work hard, play hard. I visited many nice and exotic places in those days. The most striking memory goes back to 2003 when Zara organized a regional water management conference in Cairo, Egypt. She managed not only to get this funded and organized with Shell staff but also attracted many specialists from the Arab region. We had an amazing week in the Marriott Hotel overlooking the river Nile.

Zara had this very powerful mix of Lebanese tough negotiation skills and the "can do" mentality of the USA. She also fit remarkably well in the Dutch culture of non-nonsense thinking. She liked very much the debate and you could be easily overpowered if you couldn't get back with counter arguments. I think she loved the confrontation and if you could match up with her, you had a great time. She also had a vision. That vision was materialized in the Water to Value team. Zara stood for her vision and stood for her team. Promises were real promises and she would defend them to the highest manager.

Ngadirah Kneefel, Shell colleague, Water to Value team member, and personal friend:

Zara was a "one of a kind" person, an extremely energetic positive person with a very strong character, and a mentor. My example of a power[ful] woman and I learned to do some things over the top. She was very passionate about her work and expected a lot from herself and from the people around her. She kept all team members very motivated to think out of the box, always think big, the sky is the limit, nothing is impossible, try to achieve it and give 200% of yourself.

At the same time, she was very good at motivating her people and once in a while, I was asked to organize great "away days" for the team, like a typical Dutch weekend in the south of Holland in a castle, where we had a high tea in a steam train. Or a sporting tour in a cave followed by fanatic racing in karts and ending in spontaneous dancing in the castle with all our colleagues and their families. Or an "away day" on a sail boat in Volendam and Monnickendam, sitting at the deck and seeing the local fireworks. The team, living around the world, is still talking about those trips back in the days.

I have a lot of really special memories, which I keep in my heart and Zara was at the end a key person in my life, who really inspired me. I learned from her to feel "alive" and try to dream big and make your dreams reality. Don't let boundaries stop you. She was a very lively person with a mission in life and I am very proud of what she achieved. I started to work with her as her team assistant in the Water to Value team in 2001. It has been more than 12 years since the W2V team members were all together and we are still connected by our experiences because of our Zara.

After she moved to Dubai, she always made time on her short trips to the Netherlands. The phone would ring and she would only say, "Hi, it's me," expecting me to drop everything in my agenda to see her shortly for dinner at Sakura, and I was always interested. We were invited to her villa in Dubai and had a wonderful local travel experience. I still keep my pink poodle on my desk (funniest present and it makes me smile when I see it). When I told her once it's not necessary to buy gifts all the time, I found at my desk not one gift but a bag full of gifts, not for me, but for my baby son with a note "Love, Zara," and I did the same [for her].

She organized a bicycle tour once for all team members, but I knew cycling was out of her comfort zone. I didn't accept a "no" from her and arranged a tandem and she cycled together with my husband and finished the tour. I miss the out-of-the-blue "hi, it's me" phone calls from her, her funny gifts, and her strong woman-to-woman advice.

Pascal Hos, former Shell colleague:

Zara was an important person in my professional career. She is the one that got me started in Shell. She took a big chance when she hired me when I had absolutely no experience in the oil and gas industry. At the time she seemed sometimes too driven and unfair in her judgment; however, looking back from where I am now, I can fully understand her motives and have implemented quite a number of her tactics and definitely tried to copy her passion for the job and her willingness to go against the grain to achieve the things she believed in.

I managed to dig up a photo of one of our away days in Limburg. I think it shows Zara in her natural environment, happily dancing with her daughter surrounded by a team of people that meant everything to her.

Lih-Der Tee, Shell colleague, Water to Value team member, and personal friend:

Zara's warm personality and generous heart have touched many. She saw opportunities instead of challenges; potential instead of inexperience. We are forever thankful for the opportunities she opened to us. At work as well as after work, Zara was truthful with what she envisioned, thought, said, and did. Most of Zara's team members continue to stay connected in one way or another after the team had been disbanded many years. To me, Water to Value was not only a working team that she formed at Shell but a magnet that continues to pull us together. It is the leadership quality that remains relevant till today. Cherished memories of the time we spent together continue to bring warmth to our hearts.

Ray Lesoon, Shell colleague and personal friend:

Dr. Zara Khatib was my supervisor, mentor, and a personal friend for nearly 25 years. We struck a world-class alliance from the first project we worked on together by using her unbelievable talent to look at Process Instrumentation Drawings (P&ID). Zara had the ability to recognize inherent mistakes with equipment and design layouts and proved this by her extended results with patents and solutions. Dr. Z.I. Khatib was known throughout the "Shell world" for delivering solutions and remedies for petro-chemical plants (downstream), pipelines (midstream), and offshore platforms (upstream).

While working in the offshore arena, there was a need to monitor oil water discharges for environmental compliance within the industry. I can remember one specific incident. There was a drilling mud spill in the Gulf of Mexico that produced an oil sheen. To protect her company's assets, we were chartered with developing a method and sampling technique to detect the source of the oil spill. Through this project, there was a patent that evolved by creating a sheen sampler. This process through analytical methods gave an oil fingerprint to identify the source of oil from spills, drilling mud or natural oil seeps, and continues to be used offshore to this day.

Through numerous Shell projects, involving research and development, sustainable development, and environmental compliance, we have traveled around the world to most of Shell's operating assets – Brunei, Malaysia, Oman, Dubai, the Netherlands, and the North Sea. Investigating fugitive emissions and other environmental concerns that occurred, we were able to help and ensure environmental compliance [in] third-world countries by performing an operational audit on compressors in the field. Entrained oil in the gas discharge of these compressors caused large plumes of dark black smoke being flared. This was an indicator that the primary separation was not working properly. Through Dr. Khatib's efforts and innovative mind, she created another patent that was produced to decrease the amount of carry-over in gas streams. Again, this procedure is being used around the world today and marketed by a leading chemical company.

One of Dr. Zara Khatib's strongest traits and probably one that gets lost within the corporate world is the fact that she had the ability to identify someone's strengths and pull them out to be used while working with her not FOR her. She mentored me and taught me full force all about water treatment and oil and water monitoring which included offshore as well as onshore. This led to many industry group memberships including SPE (Society Petroleum Engineers), API (American Petroleum Institute), and PERF (Petroleum Environmental Research Forum), which led me to be a Board Member on the Produced Water Society for over 25 years. She was invaluable in making one feel as an equal and this resulted in making me feel that I was always to do my very best. Dr. Khatib was a teacher at heart, she always took the time to explain the total process, never suggesting a quick fix.

On the lighter side of life, Zara and I had traveled from Houston to Tokyo, Singapore, Brunei, and finally ended up in Malaysia for a conference in Miri. Hungry and a bit lost, we arrived at Apollo Seafood, an open-air restaurant. We were informed that we had to have cash only for our meals. We immediately hopped back in the rental car in search for an ATM machine. There were many banks in Miri, but few ATMs were working on a Sunday evening. After getting in and out of the car several times, Zara drove across the street but, before exiting the car, a motorcycle officer stopped us and informed us Zara was not wearing her seatbelt. I interrupted the conversation by handing the policeman my US passport and told him we were from Houston. He responded that his family had visited Houston and, as he recalled, it was a law that everyone in Houston wore their seatbelts while driving. Therefore, he proceeded to lecture us – with no ticket to our surprise – and gave us a police escort to the hotel. On the way, he and another motorcycle cop broke many traffic laws, running red lights, going down one-way streets, etc. When we finally had dinner at the Holiday Inn that evening, we had quite a laugh and many stories to take back to Houston.

Caroline Legg, Zara Khatib's daughter:

I think what made Mom special is where she came from and how she got to where she was. Her mother was married to a cousin at 12 years old and was determined rather that her daughters would be educated. "No consideration of marriage until after your first degree!" Zara was born with a strong work ethic and drive to succeed and you need that to overcome the traditional pressure to just do the minimum to get by. She didn't care what grades she got as long as she was the best in the class. She was the first girl to go to a technical high school in Lebanon. Even then, as the only girl she would always get the top grades. Lebanon became unstable in the '70s and an Israeli bomb exploded next to her giving her shrapnel in her legs. Consequently, she would never accompany my dad and me to air shows.

She did her Bachelor in Chemistry and had almost completed her master's when the American University of Beirut was closed due to shelling during the civil war. She then went to the UK to do her master's and PhD in chemical engineering. This made it possible for her family to get out of Lebanon as the civil war worsened. Eventually, when she went to the US with her first marriage, she took everyone with her again. Her parents and sisters were able to move out of Lebanon and become US citizens because she kept looking forward.

Having two working parents meant I cycled through nannies until I started calling one of them "Mama." Then the family decided that my grandfather would move in with us and he became my nanny. To be fair, our family had more of a role reversal. Mom was very ambitious as a starting engineer in Shell. She wanted to do everything that women typically weren't doing before to show she could and she'd be good at it. My father was older and established with his career. He liked taking care of me and was more involved in activities and "raising me." Work didn't end at 5 PM for Mom; it usually kept going quite late.

I still remember her first assignment involving going offshore was in July of 1996 for a month for the startup of Shell's Mars deepwater platform, where she had invented the new oil-water separator (her third of many patents).[62] I'm sure it was so exciting because [there still were not many] women who had gone offshore. But when she got back, I didn't remember her and that was a big shock for her. It's one of the first things I remember clearly. Her coming home and wanting me to sit on her lap and hug her and I wouldn't get off Dad's lap because she felt like a stranger. A month is a long time for a kid.

When Mom was not involved with work she often worked on Society of Petroleum Engineers (SPE) papers or created other presentations. She was an SPE distinguished lecturer twice, visiting dozens of countries around the world. *I must be clear, I am very proud of what Mom accomplished.* And the role reversal worked for us. Dad was my Girl Scout leader and Mom eventually made the majority of the money. So [the way] a typical girl would feel towards her dad was how I felt towards my mom.

Were there benefits to having a mom in the oil industry? We traveled a lot. Almost every vacation we went somewhere new and there were conferences in fancy hotels and restaurants. I got to do a lot of traveling and was exposed to more interesting people and things than most kids.

Mom had a lot of success with her Shell assignment in Holland and was happy at work. However, she was so unhappy with her work experience in the Middle East that she would tell her friends to not work as hard as she did, that there are more important things in life like family and enjoying life. Five to 10 years before, she was pushing people to do more and work more to achieve. Her last couple of years she just wanted to spend time with me and we finally got very close.

She might not have had a lot to do with my life growing up, but she affected the lives of a lot of engineers, male and female, young and old. She was inspiring with all of her speeches and presentations, including twice to the United Nations. And she was a problem solver. She could pull the potential out of anyone.

If I were going to describe Mom in a sentence, it would be: *She kept moving forward.*

♀ ♀ ♀

---

[62] "The patent for Zara's Distributor for Liquid Separator is No. 5,458,777 "It allowed for a smaller vessel from existing vessels with improved separation with less plugging and was not sensitive to the distributor becoming unlevel, all important since it was being mounted on a floating tension leg platform," explains Corbett Legg.

15	## Las Mujeres en las Plataformas de Pemex
	## (The Women on the Platforms of Pemex)

> "We build barriers and inequality as people and,
> [just] as we build them, we must overthrow them."
> **Engineer Mary Betanzos Espinoza, Technical Specialist B**

These are exciting times to be in the energy industry in Mexico. In December of 2013, then-President Peña Nieto signed into constitutional law energy reforms that ended the 75-year monopoly of the state-owned oil company, Petróleos Mexicanos, known as Pemex, opening the Mexican energy market to competition. With the high-profile appointment of Mexico's current Minister of Energy, Rocío Nahle, a chemical engineer with a petrochemical specialization, who spent her professional career at Pemex, the new government headed by President Lopez Obrador has shown its confidence in the ability of a woman to oversee a traditionally male-dominated sector.

Inclusion Management in Pemex explains the company has been working in partnership with the United Nations Development Program (UNDP) in Mexico since 2014 to design and implement an Institutional Strategy of Social Inclusion, which contains guidelines pertaining to gender equality. One of the joint projects focuses on women who work in technical fields like the sciences, technology, engineering, and math (STEM) and features short online videos of the women talking about their careers.

Pemex identified three female employees, all of whom work a 14/14 rotation schedule offshore in Paraíso, Tabasco, in southern Mexico on the Gulf of Mexico, where the Dos Bocas port and terminal are located, and the site of Mexico's first known oil discovery in 1863. Inclusion Management recognizes them as "very talented employees," but also refers to them as "equality pioneers."

Not only are the women defying gender stereotypes, but their families have defied cultural stereotypes. Chemical engineer María Franco Ramón cites her father as the biggest influence on her career. Engineer Mary Betanzos Espinoza had an older brother, also an engineer, who worked on the offshore platforms in the

Photo Credit: Pemex

Bay of Campeche and, on his days off, he would talk to her about his work at sea and suggested it would be a good place for her to develop as a chemical engineer. Although she had previously worked in the downstream sector of the oil industry in refining, she viewed working directly where the hydrocarbons are extracted "as something interesting and a great challenge."

Mary says it is this brother who has had the greatest influence on her career. "He was the one who taught me to think as a professional, to set goals and achieve them without conceiving obstacles."

Although Brenda Medina Ávila, a petroleum engineer and specialist in well operations, says she has had supportive bosses throughout her career, and there isn't one specific person she would credit as being the most influential, she does mention a male production manager she encountered early in her career, who taught her a lot and impressed her with his passion for his work.

Inclusion Management calls engineer Olga Cantú Rodríguez "an important leader and ally in promoting gender equality in Pemex Exploración y Producción (PEP)."[63] While in her capacity as Manager of Programming & Evaluation, Subdirection of Production Shallow Water Blocks AS02, Olga doesn't have to work offshore, it is her belief that the managers and the women's male colleagues have a profound effect on their careers in the sense that they "drive us to develop our skills."

In 2016, industrial psychologist Carla Ivón Soriana Roldán, 34, and a married mother of two, was one of only six women, along with 214

---

[63] The Exploration & Production (E & P) subsidiary of Mexican national oil company Pemex.

men, working on Pemex's fleet of offshore rigs. In an interview with *Agencia EFE*, she said her male colleagues treated her with paternal fondness (no byline, 2016).

"The risk comes from the materials," she was quoted as saying. "The majority of the men are very respectful."

The other women concur. "At the beginning, they treated me as if I were the daughter of everyone," says Brenda, 32, and a mother of one child. When she began working for Pemex seven years ago, "I was the smallest in both age and height on the platform so everyone took care of me. When I learned the whole process, they took into account my opinions and contributions, and soon I began to give instructions to the staff I was in charge of, and there was always cordial and respectful treatment."

Olga goes so far as to point out, "Respect for professional colleagues is one of the values mentioned in *Plataformas Marinas*.[64] [It states] in the regulations that any lack of respect for female staff entails a rescission of the contract."

Mary, a 47-year old single mother of one, who has been employed with Pemex for 16 years, has been with the company long enough to remember when attitudes toward women working offshore were different than they are now.

"Older people – manual staff – were surprised because, at that time, there were not many Pemex women aboard, except one or two, a doctor or hotel manager, but not engineers and some treated me with much respect." However, they were curious what her parents thought about her decision to work at sea and be away from home for so long in a job they viewed as "very dangerous for a woman."

Men offshore everywhere were having to adjust to women working with them and Mary says they were not always eager to embrace change. Their reactions varied from telling her "this place was not for me [because] they saw me very vulnerable," to some who treated her like their sister and "[were] sent to take care of me when I was in the industrial area," to yet another extreme – "There were even those who asked [that I] not to be allowed to go on platforms!" At times, she felt there was professional jealousy on the part of her male colleagues and she even had bosses monitor her behavior and performance, asking her colleagues how she was doing, and how they treated her.

---

[64] Translation: Marine (offshore) platforms

Back then, Mary says, "There were times that I was the only woman with more than 300 men!" She only had the occasional female doctor or a woman company engineer, who carried out activities onboard, as companions. While there are more women working offshore today, based on her own observations, Mary estimates that of every 15 male engineers, there are only three female engineers, and says despite more openness within the company, even now few women hold the top leadership positions.

Brenda says things have improved somewhat in recent years and that now on a platform of 200 people, there might be 180 men and 20 women. "I believe that being a woman and working in this industry [are] not an obstacle but rather a challenge. We must remove the stigma that engineering is only for men, it is also for women... Here I am!"

Olga stresses, "Talent development is very important," and explains that a mentor is designated for the selected personnel for one to two years, after which time the employees are assigned to an area where they can better develop their skills. Before working on offshore platforms, the initial training, including well drilling and production maintenance, is conducted onshore. The next step is to work on the offshore platforms, along with personnel experienced in a particular discipline.

Female role models have been scarce for women like Mary and María, who are approaching two decades with the company and who, like petroleum geologist Paty Ortiz Gómez, who began her 41-year career with Pemex in 1975 (Gries, 2017), are among the pioneering women offshore. María simply says she looks to God for guidance. Mary, who had the benefit of her older brother's insight, says, "I have been fortunate to have good leaders in my professional development who have given me their support and guidance to reach my goals during the 16 years that I have worked at the company." Brenda echoes that, saying she has had support throughout her career "from the manual staff to my bosses [with each] contributing to my growth as a professional."

The videos that Pemex has filmed in partnership with the UNDP under the banner, *For A Mexico With More Scientists, Engineers, and Mathematicians,*[65] will provide virtual role models for young women considering a career in the petroleum sector, as well as those currently employed in Mexico's oil and gas industry, as they see and hear women talking about their careers, the challenges they have faced, and how they have navigated their way to success in a male-dominated industry.

---

[65] Videos are the copyright property of Pemex and UNDP.

Petroleum engineer Alejandra Hernández, featured in one of the videos, and an expert in the design and completion of oil wells, says she was an inquisitive child who dreamed of becoming an astronaut. Instead she went on to study engineering and, in her first semester of university, out of 25 students, there were four women, and she was the only petroleum engineering student. She muses whether this was because it was "the kind of career that was thought to be more for the male, not so much for the women."

Alejandra believes that there are opportunities for women to compete at any level. "The only thing that we require is willpower, to have clear objectives. It's not where life takes me, it's where I want to go, and I must fight to get there. Maybe we are still a limited number of women, but we have achieved [this goal]."

As women throughout Mexico's oil and gas industry become more visible, they are coming together to support each other. The Women's Energy Network (WEN) opened its first international chapter in Mexico City in 2017 under the leadership of founding president, María Luisa Licón. The organization Women Offshore, founded by chief mate Ally Cedeno, a former senior dynamic positioning operator now studying for an MBA at Rice University, is open to women around the world who work on the water. In addition to its online platform and resource center, it offers a virtual, peer mentoring program called MentorSHIP.

For young women considering a career at sea, María recommends not overthinking it, instead doing what she did 18 years ago – if an opportunity presents itself, "Take it!" Not only does she believe they can learn many facets of the industry, enabling them to grow and advance in their careers, she calls working offshore, surrounded by the water of the Gulf of Mexico, "A beautiful experience."

Arlén Garrido Díaz, a geophysical engineer and specialist in seismic attributes, makes an important point in one of the Pemex/UNDP videos when she says, "Working on platforms is not for anyone. There are physical and environmental limitations... but it depends on the person not on the gender."

In another video, geophysical engineer Rocío Negrete Cadena, an expert in seismic velocities, says, "For young women, who desire to study science [or] any career that is stereotyped as only for men, I can say that those are for anyone. The only thing that you need is to [assume] a lot of responsibility. In a world of men, in a difficult world, in a complicated country like Mexico, it is necessary to search for opportunities."

Mary believes they will find them. "The opportunities are for all, men and women; there are no special professions for each. With preparation, effort, and desire for what we do in any profession as a woman, we can be successful and achieve the objectives that are proposed. We build barriers and inequality as people and [just] as we build them, we must overthrow them."

16	**Mieko Mahi** **Photographer/Videographer**

**"I don't think I could have made it in my
career if I had pulled the feminine card."**

     I had a Polaroid camera and I knew I wanted to be a photographer," Mieko Mahi says resolutely. "I was 11." As a young child, she would take cereal boxes, turn them over, pretend she was photographing the cereal, creating the wording and selecting the colors on the box. As she got older, she would spend hours on end analyzing magazines, inspired by the photographs she saw in *National Geographic*. "I pretended to be the photographer on every single image, imagining holding the camera, staging and taking the shot. I did this on every photo of every magazine."

Born in Tokyo, where her father, Captain William Rufus Hass, was stationed in the Air Force, Mieko's mother, Dorothy Louise Allen Hass – who embraced the beauty and culture of Japanese life: kimonos, music, porcelain and ceramics, art, the green tea ceremony, luscious cherry blossoms, Ikebana and Bonkei[66] – gave her friend the honor of naming her youngest daughter, whose name means "beautiful city." She later found having an unusual name worked to her professional advantage, as it is not easily forgotten, particularly one that is gender-neutral in a male-dominated industry.

The family would move to Alaska when Mieko was five and stay until she was 11. She doesn't recall being aware of the oil and gas industry, but instead remembers it as a magical place with Northern Lights and huge billowy snowdrifts. While it was one of their longer

---

[66] Ikebana is the ancient Japanese art of flower arranging and Bonkei is the Japanese art of creating three-dimensional miniature landscapes in a shallow tray.

postings as a military family, the reason the memories of that time are so vivid is because her father was still alive.

"I was a Daddy's girl. I am like him in that he was a painter and a writer and we would always go to the movies together. The fun was choosing one at random and the surprise of finding out what it was about. He lived life to the fullest – he was rarely seen

without a cigarette in his hand, cussed in his everyday speech – although never toward his family or in anger – drank whiskey, favored spicy foods, and worked long hours. He was a glider pilot in WWII and was in D-Day. He is quoted in Cornelius Ryan's *The Longest Day: The Classic Epic of D-Day, June 6, 1944* (1959) and listed among the D-Day veterans.[67] He was a hero."

The family moved to Detroit, in the northeastern United States – in the aftermath of the "long hot summer" race riots of 1967 – where Mieko's dad worked for the Federal Aviation Administration (FAA) as an air traffic controller until he suffered a fatal heart attack when Mieko was 15.

The youngest in a family of four girls, "I learned everything the hard way because I didn't have anyone holding my hand on my college or job search. My dad died and my mom became a widow."

Longing to get back to the things she enjoyed, such as flower arranging and volunteering her time, Mieko's mother moved the girls to San Antonio, Texas. "I left home at 15 – without my mother's blessing, although she did send me $60 a month from my dad's Social Security – moved to Dallas, and put myself through my senior year of high school (the fourth in three years) walking to and from school and work. I lived in a garage apartment for $75 month. I'll never forget I shopped for groceries on $15 a week or maybe less. That's why I refuse to look at prices now, because I had to go without for so long."

It was also during that time that she used her brand new credit card to buy her first camera, an Olympus OM1 ("I loved that camera"), and

---

[67] Hass, William R., Jr., F1/Off. [441st Troop Carrier Group], Capt. USAF.

her older sister, Joan, paid for her first photography class, enabling her to pursue the dream she had had when she was just 11.

In college, she studied photography and art, and a broad range of media – radio, television, and film – later incorporating advertising layout and computer science, and graduated from the University of Texas at Dallas in 1988 with a Bachelor of Arts, Visual Arts, although not before she went through some very trying times.

For many students, an internship is a golden ticket into the working world. Mieko instead learned a bitter lesson when the attorney she reported to at the film production company where she was interning asked her to turn over the negatives to her film. Mieko refused, telling him she owned the copyright.

"He showed me the door and I got a C for my internship. That's when I learned being right isn't necessarily winning. I thought I would never work again – ever – and that word would get around I worked for free and got fired. I was completely devastated. I can still see myself lying in the bathtub crying for hours."

Mieko believes it wasn't only the power struggle over the copyright that caused her to be terminated from the internship. "I asked really good business questions, and that movie production company didn't want to tell me one thing about the way they handled their business. I was supposed to act dumb and not ask anything, which is probably good advice when breaking into a career... not to act dumb, but to be quiet and watch and listen and read instead of just asking questions. Even more important, you have to experience it, do it yourself."

Leaving college temporarily, Mieko says, "The real problem was that I expected too much of myself and I was taking on too much. I wanted my journalism to be published along with my photographs that I printed in the darkroom myself. I wasn't interested in just doing the work at hand like the other college students." Ultimately, it became over-whelming and she says, "I had to get away."

Shortly thereafter, she landed her first corporate photo shoot for Southwestern Bell Telephone Company. The job came at such a critical time in her life, she still has the check.

Despite money and resources being in short supply, Mieko joined forces with her friend, photographer Margaret Kershaw, whom she had met in a darkroom when she was attending North Texas State University in Dallas, and the two went into business together. The young women became very close and partnered on fashion and product assignments, honing their photography skills. Never sleeping more than a few hours at a time, working and networking as much as possible, Mieko recalls that time in her life as "carefree" and full of creative energy even though she struggled to make ends meet.

"Imagine being a photographer in the '80s and staying out from sunrise to sunset with fashion models and business people. It was a blast! It was a really cool roller coaster ride."

Having discovered the highly-competitive nature of creative work, Mieko worked hard to become established in her chosen field. "I had to train myself on the job." Mieko went to an interview for the job she calls her "first big video career break" in a borrowed suit jacket and rundown shoes. Recognizing her potential – and overlooking the holes in her stockings – film director Malcolm Neal hired her over the other applicants.

For the next three and a half years, she would work on location and in the studio as a producer and director of orientation, educational, and training videos, and then for nearly two years she partnered with Ruby Johnson as co-owner of a documentary filmmaking venture.

At the end of 1988, she was hired by Texas Oil & Gas Corp. (TXO) to create a one-woman video department based on her ability to do it all – shoot still photos, write, edit, produce and direct video, set up lights, regulate sound, and operate the video camera. However, making light of her skills and talent, she told the Public Affairs Manager, Mike Dixon, that she could shoot "party Polaroids." Taking her at her word, he hired her to take photos at a 50s-themed company party. "I showed up in a poodle skirt and the rest is history" (Economides, 2013).

When TXO sent her to an offshore platform, Mieko had no idea what to expect – including the fact that seasickness is one of the hazards of the business, as many offshore workers can attest. She says the men onboard made it worse by laughing and talking about food and other unpleasant things, but she had her revenge. "I worked the whole day, taking photographs in between throwing up. I didn't throw in the towel and pass out like I have seen other people do since then. I remember asking politely between photo shoots, 'Excuse me, Captain, may I throw up over the side of the boat?'"

Getting seasick paled in comparison to the realization she had on her first helicopter shoot. While hovering about 1,500 feet in the air, "I leaned out to get the shot. Afterwards, I looked down and all I had on was my seatbelt. I never, ever did that again." In an understatement, she adds, "To get aerial shots, it's better to be tethered, to have that second safety apparatus.

"Then there was the time I overdosed on seasick patches." She hadn't read the directions and took off one patch and put on another one the next day. The sweltering heat didn't help. "It was 100 degrees out. My close-up eyesight went out. I couldn't read the notches, the controls on my camera. I'm sick and I can only see large objects. The show has to go on. I'm sitting on the back of the boat to see the horizon, crackers in one hand, 7-Up in the other. I asked the crew to help me. I told them how to set the camera. They talked to the captain, they coordinated the whole shoot. I literally stood up and took the picture and went back down." Amazingly, she was satisfied with the end result, especially she says because it showed cooperation.

After nearly two years, Mieko moved to Marathon Oil as senior corporate photographer/videographer, where she was the photographer for the company's annual report, stockholder and employee publications, and produced turnkey videos for Human Resources and Legal, all while continuing to go offshore.

In 1994, with six years of experience in the industry, she decided to strike out on her own as a freelancer with encouragement from her new-found friend, the late Richard Payne, who would later become her mentor. She met Richard, a renowned architectural master photographer, after calling to inquire about purchasing one of his photos. He had refused an offer from a major oil company and Mieko was impressed by the value he placed on his work. His advice was simple: he told her if she did what she said she was going to do and charged reasonable prices, she would be successful.

"I love Richard Payne and respect him so much for his love of the fine art aspect of photography," Mieko says of her friend, who shared his insights right up to his passing in 2018. To this day, his words of wisdom continue to be her guiding principle and business motto: "I do what I say I am going to do."

Jay Brousseau, whom she refers to as a "big-time Dallas photographer," has been a mentor since she began her freelance career and someone she now counts as one of her best friends. Early in her freelance career, Jay offered tips on what she calls "must-know lighting techniques" and, 30 years on, continues to share his knowledge and expertise when Mieko finds herself with a particularly challenging assignment.

There were two women whose belief in her made a lasting impact on how she saw herself. The late Jean M. Longwith, the chairman of the radio, TV & film department at San Antonio College, one of the schools

Mieko attended, wrote referrals for her, and later in her career the late Marilyn "Maren" Williams Linley, who was with Public Television Channel 8, sent her so many leads for full-time jobs Mieko finally had to ask her to stop as she didn't want the temptation to leave freelancing. She refers to the two women as "the encouragers."

When she first started freelancing, Mieko estimates she would make 45 cold calls a day, something that was a lot of work in itself, but she was determined not to fail.

"I would do two, three, five interviews a day. Sometimes, I didn't even know who I was meeting. Men would take me throughout the entire facility, introduce me to everyone, take me to lunch, have me give a presentation – and then they wouldn't give me the job. I wanted to be hired for my work, not what I looked like." She made the decision to stop doing interviews and was hired based solely on her reputation and the quality of her work.

Mieko says she learned early on not to become too familiar or share personal details with people because there is a risk of developing an attraction. "You cannot freelance and do that or you lose the account. It goes both ways for men or women."

Another lesson she learned at the start of her freelance career goes against conventional wisdom. As she emptied her briefcase one day, out poured hundreds of business cards she had collected while networking. To her amazement, the majority of the cards were from photo labs, paper companies, and other photographers – some of which were her competitors.

"As much as I enjoyed meeting these people and getting to know them up close, I stopped going to those events because time, energy, and resources are limited." Instead, she found her time was better spent learning about new technology, listening attentively at business meetings, determining how the information related to her, and how she could be a better photographer and communicator.

On her very first freelance job offshore, Mieko was on a workboat and wanted it to go as fast as possible. Fearing she would fall overboard, she had the deckhand hold onto her shoulder. She heard "this squeaky sound" that was actually the captain yelling that they were about to be engulfed by a wave.

"I had a split second to make a decision. I took off my shirt and covered my (very expensive) rented video camera. I lost the account because water got inside the lens and damaged the camera. I didn't get paid. And I learned to always wear an undershirt! I had to decide what was more important – being appropriate or saving the camera. Some people would have quit over an incident like that. I had to go out and get a job the next day. I was determined to stay in a career where I had proven myself. It was a gutsy thing to overcome."

Perhaps it helped set the tone for her career going forward. She didn't want special attention and says she did as much as she could to be like the men, explaining that she was influenced by her military upbringing.

"I felt like I was a guest in their environment. I never wanted to change them and I certainly didn't want to be treated differently because I was a woman. I tried to hide my femininity and let my work speak for itself." At one point, she wore the same baggy clothes for 10 years and went so far as to wear men's underwear when she was offshore, so that when it was put in the laundry it all looked the same. "I don't think I could have made it in my career if I had pulled the feminine card. I don't think any CEO would hire someone who would be a distraction out on the rig."

Even a major life event, like marrying husband Darrell Wachel in 1997, was not allowed to detract from her work. "The thing about the oil patch is scheduling: it's never on schedule. It's always guesswork because a lot of factors go into a launch date – setting and meeting deadlines for getting the personnel, transport, and equipment scheduled.

"At my bachelorette party, I had to have my steel-toe boots and hard hat in the trunk of my car and watch how much I was drinking because I was expecting to be called offshore any moment. Well, it didn't happen. It got pushed back and back and back, and then it was my wedding day.

"I told the marketing sales manager, 'I'm getting married tomorrow,' and he said, 'Make up your mind, Mieko — do you want to do the shoot or not? I can call someone else but I'd rather have you.'

"I said, 'I'll be there late afternoon and I am bringing an assistant to carry my gear and, oh by the way, I'm marrying him.'

"And he replied, 'I am not paying for a vacation; you have a job to do!'"

Having gotten married that morning, Mieko and her new husband had lunch with family and friends at their new home, then changed clothes in the limo as they were whisked to the heliport. "In this biz, you don't turn down work; you'll be off the list. People stop calling," Mieko says matter-of-factly.

When she and her "assistant" got to the rig, she says, "The fellows were not exactly friendly; they almost didn't let us have the same room." She assured them they were in fact married and, much to her surprise, they were given an "unusually nice room with faux marble in the bath-room and larger than normal bunk beds." The next morning, by the time her husband woke up, Mieko was already done with her shoot.

"My husband is not in the oil business, so he had an eye-opening experience seeing me on a rig with 100 men." The newlyweds then made the three and a half hour flight back to shore, where the saga continued when their ride didn't show up. They ran into one of Mieko's girlfriends and resumed their celebration.

"We ended up flat on our faces at our house that night, but I got the film to the photo lab."

Three years later, even the birth of her son, Reese, didn't slow her down for long. She was back at work in 30 days, although it was an adjustment. "I had to realize that I did not have the freedom to sleep, eat, work when I wanted to any longer, I had to *manage*." It helped that both she and her husband worked from home and were able to hire a nanny. On the rare occasion that she found it necessary to show up for meetings with her son in tow, she discovered that clients were considerate and understanding. As much as she enjoyed motherhood, it didn't decrease the rigorous pace of her work schedule.

From the Amazon to locations on the coast of Africa, Mieko's career has resembled a real-life action adventure film. After being flown by private plane to Peru, she once spent 22 days in the jungle on a com-pound surrounded by chain link fence and guards patrolling for anacondas and jaguars. She has "fond memories" of ants crawling on her face and video camera while she tried to stay still long enough to get good footage of cargo coming in at a refinery expansion.

Given the physical nature of the type of photography Mieko does, it is crucial for her to be in a creative mindset, remain focused on the assignment, and not be distracted by fatigue or hunger.

"I have an unusual talent. I can sleep anywhere – in an airport, a boat, a chair, an aircraft – and in any position. I don't have any pride when it comes to sleep because I know I have to rest." Not being able to eat on time during an assignment is a pet peeve and isn't particularly about the food. "I want the break to rest, clear my head, and stay nour-ished so I can put in an eight to 16-hour day without any problems."

Mieko later returned to the Amazon on a 14-day trip, creating video and still images of cargo being unloaded from a large cargo ship to barges that would deliver immense towers and heavy crates to a construction site. While she would have preferred to stay on the vessel, she had to travel back to her hotel alone at night, unable to speak the language, in a motorized rickshaw.

"Safety is constantly on my mind," Mieko says, expressing her respect for safety managers and her appreciation for the briefings they hold to reiterate the safety requirements. "That's not just from a woman's perspective; I know for a fact men feel that way, too."

There was the time she received a somewhat mysterious phone call to shoot a nuclear energy plant in Mexico and was told to fly to a certain location. "Despite a part of me thinking, 'Who is this guy, who is this company?,' it worked out. I was mostly worried as I waited for him to pick me up and was relieved when he showed up in a black Porsche SUV. It was legitimate; I got paid."

Then there is what she still animatedly refers to as "the most exciting trip of my life!" Upon arriving at Campeche Sound, in the southern Gulf of Mexico, to conduct seismic photographing of gunboats shooting vibrations, she was placed under house arrest when it was discovered her visa had expired. After some negotiating, she was allowed to board the helicopter, which then went out of control and began swaying from side to side, almost crashing.

"The pilot's freaking out. I'm in the front and somehow we land. Do you know what it feels like to have to get back on the same helicopter? Nerve-wracking!" she says, answering her own question. "I had to get back on the craft within minutes of evacuating and I wasn't so sure it could actually get off the ground. We landed on the helideck of a large offshore vessel. Then we had to change vessels and wait until the boats came together and jump from one to the other. Timing is imperative!

"Once we got back to land and the helicopter touched down, a belligerent drunk guy was fighting with the police. I was afraid he would get shot. The excitement never ended," she says wryly.

Somewhat less adrenaline inducing was the assignment that required Mieko to be offshore for 18 days – the longest amount of time she had ever spent offshore. So long, in fact, she had to start deleting photos because she was running out of hard-drive space.

"I've learned to pace myself. I wanted to get on a rig one time and asked when they were going to pull the pipe out of the hole. The company man said one o'clock and that's the shot I wanted to get. He told me I had three seconds." Another time, she was told she could circle a drill ship only once. In both cases, Mieko says, "They didn't know me; that's all the time I needed. People hire me for that reason. I see myself as a problem solver not a problem causer."

Her big break came when she held an exhibit at the Ocean Star Offshore Drilling Rig and Museum in Galveston, Texas. "After that, customers would call and ask if I was available and hire me. I didn't have to bid." With photographs on the covers of such well-known industry publications as *Oil & Gas Journal, Oil & Gas World, Oil & Gas Investor, Petroleum Engineer International, Hart's Pipeline Digest, Business Week, Offshore Magazine,* and *WorkBoat Magazine,* as well as in the *Houston Chronicle* Business Section, a *National Geographic* textbook, and a *New York Times* feature story, she says, "At a certain point in your career, it's a trust issue. Most of my career I have worked without a contract."

Along the way, she would meet the late renowned energy analyst Michael Economides and become friends with him. She learned some valuable lessons from him just through observation.

"He had a unique confidence as if he couldn't wait to speak to a crowd and share what he knew. That confidence was appealing because it's so unnatural for most of us. He was energetic and didn't hold back. I have a girlfriend that is a good public speaker and I can see her hold herself back on purpose for fear of appearing overly confident to other people, but Michael was a person that displayed his confidence without the slightest concern that someone might not agree with what he was saying."

He showed his confidence in Mieko's talent and skill as a photographer by always asking her to provide what she thought were her best photos to use in his books.

In a profile of Mieko, Michael wrote, "She is a person with a mission: to show the goodness of an industry that has been savaged by bad publicity" (Economides, 2013). She expands upon that by saying her work shows the brilliance in the technology – which she views as an art form – with dynamic images.

Her aspirations are validated when someone comes up to her after a presentation or an exhibition and tells her how her work has affected them.

A man once told her, "My wife has a whole career because of you! She was fascinated by your work when she saw it and we figured we could do it. We made a career out of photographing race cars."

Mieko tries to save the written comments she receives and says the most gratifying come from someone that is struggling and says her story has given them hope.

"One time I gave an impromptu talk and told a story about something really bad that happened to me and how I overcame it, and afterwards a handsome young man came up to me, almost in tears, shaking, and said, 'I want to thank you for sharing that story; it means a lot to me.'

"It made me feel great because I wondered what people thought of me for sharing something that most probably would have quit their career over and, too, I didn't think I did a good job speaking that day," she says, recalling being gripped by a moment of panic. "I was freaking out thinking, 'Oh my gosh, all these people I know are looking at me.' The point is, I shared how I put this incident behind me and kept going forward, knowing I wasn't letting anything stop me from what I knew I wanted to do — and that's really what it takes. That young fellow must have gone through a traumatic experience and he needed to hear hope. You can never judge someone or know what they're going through just by looking at them."

Motivated by family values and traditions, Mieko and her husband made the decision about ten years ago to move from Houston to Hallettsville, a small town in central Texas, where her husband's family has cattle ranches. (His grandmother immigrated to Hallettsville from the former Czechoslovakia and his family married in the town's Sacred Heart Catholic Church.) Still within driving distance of Houston, the energy capital of the United States, Mieko now lives in one of the counties that is part of the prolific Eagle Ford Shale, site of the 2008 oil boom.

While continuing her work as an offshore photographer and videographer, which has earned her the moniker "the Annie Leibovitz of the energy industry" – accepting about three assignments a month, down from seven in the heyday – Mieko embarked on a new venture into the unknown in 2015 when she decided to open the Hallet Oak Gallery, which she registered as a non-profit the following year.

"The gallery is pure love and passion; it's not bringing in a profit, but it is helping the artists and the community. It's exciting to be a part of something that I am hoping is bigger than my imagination. The most successful thing I have ever done in my life is my "life on a rig" exhibit. I am hoping this gallery will be the biggest thing I have ever done in my life to help others and to share the part of life I truly love – and that is art."

♀ ♀ ♀

17	**Deirdre Michie** **First Female CEO Oil & Gas UK**

"As a woman, you were seen as a bit of a novelty."

❝ This was in '86 when we were going through what was quite a dramatic downturn," Deirdre Michie recalls of her entrée into the petroleum industry. Thirty years and several downturns later, including the current one which started in 2014, she finds herself at the helm of Oil & Gas UK, the "voice of the offshore oil and gas industry," looking to reassure not only today's workforce, but the up and coming generation that this continues to be a dynamic industry with a positive future.

As a young law school student at the University of Dundee in Scotland, Deirdre was inspired by her oil and gas law professor, a former British Petroleum (BP) senior counsel, who regaled students with stories of "amazing deals in far-flung places. He engendered the kind of excitement and variety and diversity about the industry, which really intrigued me." Her father had advised her to differentiate herself and she saw the opportunity to do just that in the nascent specialty of oil and gas law.

She began looking into graduate schemes in oil and gas companies and applying for a job. "This industry was in a downturn but Shell was still taking on graduates because it recognized if it didn't, it would open up a generation gap in its workforce – something that is of real concern again today.

"My graduate intake was the first time Shell had ever had an equal intake of females and males. There were 22 of us – 11 women and 11 guys – so I had a readymade network with some amazing people, some of whom became lifelong friends.

"But it wasn't until some of my colleagues and friends left that I realized the value of this network I had. That's when I started supporting Shell's Women's Network because I really appreciated it and wanted to provide the support to others that I realized I had benefitted from, albeit in an informal way."

Joining the graduate program meant that Deirdre would experience some diverse roles in her early years. Her first job was in distribution, where she was given "quite a

Photo Credit: Safehouse Habitats (Scotland) Ltd

paternal and helpful reception. As a woman, you were seen as a bit of a novelty. I remember going to one conference, though, and this would have been about a year in, walking into a room – a big conference hall – and it was 700 men and me. That was daunting. It would be at times like that when you would realize there weren't many of you.

"Because I was in marketing, which had a better although still limited gender balance, I didn't encounter overt discrimination, but I did have some more subtle experiences, which can be just as challenging. The overt you can just call in the moment and give as good as you get. The subtle stuff is more difficult to deal with."

Deirdre remembers being part of a mentoring network where mentees were assigned a mentor. Although she is a strong advocate of mentoring and feels the relationships can be very powerful, she says, "In my experience, the best schemes are where the mentee makes the call as to whom you want as a mentor, and that mentor is not pre-selected for you. I think you have to have the chemistry, the trust, to make it work well." She has had some very strong and helpful mentors both formally and informally, something she recommends.

"I like the formal structured aspect of mentoring which requires the mentee to drive it because it puts discipline into your thinking and your time management. But I also think that you can have many mentors for many different things, and I think we should be encouraging people to think about that. That's what I do now. Do I have a formal mentor at the moment? No. But I speak informally with people I respect and seek their thoughts and advice."

Deirdre also believes it is important to foster awareness that mentoring can be rewarding for both parties involved. "One of the best pieces of advice I got from one senior mentor is that he saw mentoring as a great opportunity to get a different lens on the organization. I think that's something that's very important to reinforce to mentees. Both sides gain in this."

For the next 20 years, as a self-proclaimed "Jack of all trades" in what she describes as quite a diverse career, she would move roles every three years or so – covering a variety of jobs, including distributor trade business manager, senior commercial negotiator, external affairs business advisor, communications manager, and contracting and procurement manager for Europe.

Early in her career, she married Paul de Leeuw and they began a dual-career marriage during which they had two sons, both of whom are now at university. She praises Shell for its "very generous" nine months' maternity leave policy, which ensured she had good maternity experiences.

In answer to the age-old question whether women can have it all, Deirdre believes they can – "at different times. I see young, ambitious females coming through the pipeline, but it's a matter of getting their timing right, appreciating the phase they are in, and making the most of each one.

"I would never have applied for my current role when the kids were younger. Managing the work/life balance with children at home would be very, very challenging. Because the kids are at university, I'm able to focus 100 percent on this job, and I actually think that's what it needs. It certainly has needed it for the last year as I've been on a massive learning curve."

She says she hasn't always gotten the balance right. In 2011, after spending two and a half years in contracting and procurement, something she describes as "a big job, quite full-on," she was faced with the fact that her eldest son was leaving for university the following year. After a time of soul searching and coming to the conclusion if she didn't do it then, she never would, Deirdre presented a business case for taking a seven-month sabbatical, persuading the company that it would be mutually beneficial.

She discovered people had very different reactions to her decision. "Some would ask, 'So what are you going to achieve on your sabbatical? Are you going to climb Mt. Kilimanjaro? Write a book? Solve world hunger?'" she recalls, laughing.

"Some people really struggled when I said, 'There's some charity stuff I'm going to do, and I'm on a couple of boards, but actually I want to be at home with the kids because my oldest is going off to university and I want to make sure I'm around.' Some people understood this and some people didn't. But I knew I was doing the right thing for me and my family," Deirdre says resolutely.

At the end of 2011, she went back to "a fabulous global role" as general manager strategic sourcing, based out of Aberdeen, and travelled regularly to meet with her international team and its suppliers. She remained in that position until 2014 when she led the reorganization of a multidisciplinary team to develop and deliver a revised sustainable operating model for Shell's UK upstream operations. She concluded her extensive career with Shell upon being appointed the first woman CEO of Oil & Gas UK, the highly-influential, 430-plus member trade organization, widely recognized as the "voice" of the offshore oil and gas industry.

While the job doesn't require her to go offshore, Deirdre thinks it is important to be up-to-date in at least the minimum training to ensure a good understanding of what is required for personnel working offshore.

"I do think to be credible in this job I need to understand what the requirements are. It is an extraordinary thing to go offshore and to be going along in the helicopter and suddenly these installations appear out of nowhere. The experience is amazing, and I think the shame is we can't get everybody to go and see it and touch it and feel it. A big challenge for our industry is that it is offshore and people can't see and appreciate the technology, the engineering, and the skills of the offshore workforce and its commitment to the industry. It's a challenge for us to get people to appreciate what is out there and what it's doing for us all in terms of keeping the lights on and contributing to the UK's economy through supporting thousands of skilled jobs and attracting investment."

As further evidence of this, Deirdre mentions research that was conducted in regard to people's understanding of energy. "If you're in your 50s and beyond, you understand what energy is because you remember bringing in the coal and the logs and you have experienced that it is for warmth and cooking. If you're in the 18 – 24 bracket, you only think of energy – I'm talking gas – in terms of cooking because that's the only time you see it. You don't see it when you're charging your iPhone and you don't see it when you're heating the water for your hot shower."

She believes that lack of understanding contributes to the difficulty in recruiting the next generation of workers into the industry and is committed to helping change this.

"I am a passionate advocate that this is an amazing industry still and that it will continue to be so. We're always going to need energy. Whether you're involved in an oil and gas context or renewables or nuclear, the skills are transferrable. From a UK point of view, we've got data from government that oil and gas will continue to make up 70 percent of the energy demand until the 2030s."

Both of her sons are chemical engineering majors at university, so it is obvious her passion has had some influence at home.

"However, we have to persuade the younger generation that the industry has a long-term future. Part of the challenge is retaining the confidence of people that as they go through this downturn that it is not personal. You have to be resilient, you have to look at the bigger picture; you do need to be thoughtful about what the opportunities are, to think out of the box a bit, and differentiate yourself, absolutely," Deirdre says, echoing her father's words of wisdom to her as a university student.

While she believes it is important to provide an element of reassurance that as individuals and as an industry we are going to come through the current downturn, she also has personal goals she hopes to accomplish during her tenure at Oil & Gas UK.

She says one of these goals has been to review Oil & Gas UK's role and mission and its objectives. "I want to make sure this team continues to be a leading voice, and a very effective one for this sector, recognizing the changes that are taking place around it. That's what we've been working on as a team, so that as the industry continues to change, continues to face challenges, we have an organization that understands that, is very responsive to its membership and its needs, whether it's the oil volatility or the move to a lower carbon future. So, in summary, that as an organization we are robust, sustainable, and people want to work with us and for us."

Deirdre is confident there is still a great future for the industry and those that work in it, if there is a focus on where we want to be in the decades ahead and on working cooperatively to get there. "I think it was this longer-term perspective that got me the job rather than the fact that I am a woman.

"It's about who I am, not what I am."

♀ ♀ ♀

18	**Scarlett Mummery** **Geotechnical Engineer**

"It's not the female thing; it's the age."

Travelling the world, working only half of the year and being surrounded by men – surely that sounds like the perfect job to any woman.

A grand total of just a mere 3.6% of women make up the offshore industry and statistics show that this number is decreasing. There is a complete misconception with regards to a women's place in the oil and gas industry. Every time somebody asks me what I do for a living – 'I work offshore, I'm a Geotechnical Engineer' the surprise on their face is always amusing. It is great to witness how judgmental people can be when it comes to a female working in such a male dominant industry. I know my limits, I'm a petite blonde girl that looks probably better suited to hairdressing or beauty therapy (not that there is anything wrong with that, every girl needs an amazing hairdresser) but I can assure you there is enough employed muscle onboard. I hope that this blog highlights the importance and offers women an insight and interest into a vocation they may not [have] initially thought of entering into.

I love my job, and in the relatively short time I have been working offshore it has opened my eyes to some of the most amazing places and cultures this world has to offer. The beauty of my job is that I am very unlikely to work in the same place twice. Site investigations are continuously being conducted all over the world in all continents. The world is such an incredibly beautiful and diverse place and I am completely in love with the fact that my job allows me to be one of the first people for thousands of years to see and touch a piece of rock that makes up this haven we inhabit.

From *The Offshore Blondie* blog
One site investigation at a time
October 30, 2014

Fresh out of Coventry University in England, the ink barely dry on her Bachelor of Science degree, 21-year old Scarlett Mummery (aka The Offshore Blondie), who hails from Lowestoft, on the east coast of England, landed her first offshore position as a geotechnical engineer. A year and a half into her career, excited to share her experiences – particularly with other women – she created The Offshore Blondie blog and posted her first entry.

Two and a half years later, the wonder has faded; if not for Scarlett, then definitely for the industry as a whole. It is experiencing a massive downturn, layoffs are rampant, and the price of oil has plummeted from about $90 a barrel[68] when this blog post went live to somewhere in the vicinity of $30 a barrel as of this writing. While she is realistic about the situation, she continues to be optimistic about the future in part because of words of wisdom from her grandfather.

> "I can't believe we lost you three years ago today, Granddad. I miss you and think about you every single day. 'Work or Want' the greatest life lesson you ever taught me. Wherever you are, I love you so much."
>
> Facebook 1/18/2016

"My maternal grandfather is Leslie Bertie Skitterall and 'work or want' is something he used to tell me all the time. It's the best advice from anyone and I live my life by it," Scarlett says. And, indeed, just three and a half years into her career, with the industry temporarily imploding, she is busier than ever with a variety of activities that in some way are linked to oil and gas.

Her hometown of Lowestoft, in Suffolk, on the North Sea coast, is the site of one of the last remaining "shout fish auctions," and has strong ties to the offshore industry. Scarlett says that, while the majority of young men from Lowestoft either work offshore or want to, she is not aware of any other women from her hometown that do so. There are

---

[68] Source: US Energy Information Administration (EIA)

energy companies that operate out of Lowestoft and Great Yarmouth, and she does know a few girls who work in administrative roles but, "I don't know anyone my age – 24 – or above that is in a managerial role."

Perhaps that is not surprising, given that Scarlett herself wasn't sure what she wanted to do when it came time to decide where to go for her A-levels and to choose a university. It just so happened, around that time, that her mother, June, had a young man come into the restaurant she owned "driving a flash car and a wearing a nice watch." Being the friendly woman that she is, as Scarlett says, her mother began talking to him and discovered he was a high-traffic surveyor working offshore. Later, she discussed this with Scarlett, who says she has always been "very good at geography," and she eventually made the decision to pursue a Bachelor of Science.

"It was a combined Honors course in physical geography and then there were some modules, which brought in the geology. I had the idea of going offshore and I knew I was going to do my dissertation on something related to it (she eventually settled on the topic of decommissioning – 'Rigs to Reef: Is this a decommissioning option for the North Sea') – so I did pick those modules and tailor them for that as much as I could."

While at university, she had a male professor, who had worked offshore previously, but after marrying and having children had gone into lecturing, perhaps having to adjust his career aspirations – a situation many women have found themselves in – something Scarlett believes he resented.

"To be honest, I think he was jealous. He actually wanted to go offshore, and I was the only one who wrote a dissertation on something to do with oil and gas. He was my supervisor for the dissertation because, obviously, he had the experience. I remember going to him and we were discussing topics and he said, 'So you actually want to go offshore?'

And I said, 'Yes, definitely, when I graduate.'

He said, 'Well, you haven't got a hope in hell of doing that.' You know when someone makes a comment, he may be saying it in a jovial way, but there is an underlying emphasis?

I asked, 'What do you mean by that?'

And he said, 'Oh, come on.'

He never gave an explanation as to why he had those thoughts, but I honestly think it was a personal issue with him, rather than against me. What he said was completely wrong, though."

Scarlett recalls ringing her mother afterwards. A successful entrepreneur, her mother owns businesses and investment properties, and recently was victorious in spearheading a campaign to prevent the local fish market from being displaced by a wind farm operation, believing the two can coexist. "She's happily married to my dad, but she's always said, 'Make your own money, see your way in life, and don't ever rely on a bloke,' because a lot of women do and some people get trapped. She's very strong and independent. She's portrayed that, not just to me, but to my brother Guy, as well.

"My parents are different personalities to each other. My dad, Odin, has a passion for trails bike riding and tinkering, maybe this is where I get the engineering aspect from because I think very methodically, and I think I get that trait from him. My mum always says he should have been an engineer or a mechanic. He's so accurate, precise, and thorough; where mum is a lighter personality, she's a businesswoman-type and she's very strong. He's ridiculously proud of me, I know that."

Her mother's example and her father's encouragement combined to give her the self-confidence she has today and to take comments like the one from her professor, as well as others she has heard in reference to her job, with "a pinch of salt." She also has discovered there are men who do not tolerate sexist comments directed toward women, something that was vividly illustrated by an incident that occurred during her Standards of Training, Certification & Watchkeeping (STCW)-95 course before going offshore for the first time.

The training took place in a pool and the participants were required to come out in swimwear and get into their coveralls. Scarlett, who had turned 21 that March, finished her last university exam in April, and then enrolled in the course, wore a bikini because "that's what I had. I'm not going to be a granny in a three-piece swimming suit. I don't own one and that's as simple as it was." During the fire portion of the training, a middle-aged man who was taking the course persisted in making comments about her "white bikini." While Scarlett says she laughed off his remarks, he was overheard by one of the firemen conducting the training, who then reported him. The man's employer was contacted and he was not allowed back on the premises to complete his course.

"So, other people, who are men, find these things just as disrespectful."

Safety training can be a great equalizer. Scarlett, who says, "I weigh in at almighty 48 kilos[69]" and doesn't consider herself very tall at 5' 6" (at least not in comparison to her mother who 5' 10" and her brother who is 6' 3"), saw a "big, strong boy" much larger than she pass out from heatstroke, and some of the men opting to wear different colored helmets, indicating they were not strong swimmers. "I think this is where the female thing, again, probably goes out the window," she muses.

She signed a year's contract with Gardline Geosciences, the company she credits with giving her "my step into the industry," something for which she is very grateful, and went offshore, working a month on and eight days off.

"I travelled a lot during my time at Gardline to places including the North Sea, Norway, and Gabon, West Africa. I was involved in a variety of site investigations from offshore wind farms to relief wells." Once her contract was up in May of 2014, she made the bold decision to go freelance.

While acknowledging it was a brave thing to do, she says, "To be honest with you, I was 22, had no responsibilities, I lived at home with mum, I was earning good money, I didn't have children to worry about or a mortgage, so I thought, 'What have I really got to lose?' It was just one of those things that made complete sense."

Scarlett was immediately hired by Benthic, which specializes in offshore marine geotechnical investigations, on a "casual contract." The company flew her to its Singapore office and while there she made her TV debut on *Tools and Technology Solutions Impacting Industry Success*, a one-off show, on the Discovery Channel (2014). "It is such a well-run company with such potential, and it's so exciting to be involved at this stage because I know where I want to be in five or 10 years and I think that they do, too."

She also admits her decision to go freelance was influenced by the way she has been received offshore. "The last boat I was on was in Mozambique at the start of 2015. As soon as I boarded, the client rep came up and said, 'Oh, you're a breath of fresh air being on here.'

---

[69] 105.8 pounds (one kilo = 2.2 pounds)

They're all so interested in how I got offshore that it breaks the ice, then I've made such great friendships. I think that gave me the confidence, maybe more than some of the guys, to go freelance. People say, 'You do stick out like a sore thumb,' and that's as simple as it is, but I also work hard.

"I think there's just a misconception about how women get discriminated against. When I go to work with a new crew, some men do look at [me] like, 'Oh, what good is she going to be? She's not going to be able to lift anything.' But, you prove yourself. If you work hard and you know what you're talking about, there's enough 'employed muscle'" – a term she is quick to point out a male colleague uses to refer to himself – "that will help with the more physical tasks. You're all friends, you're a team; it's not about weaknesses."

Having said that, Scarlett does feel there are times when her youth – not her sex – works against her. Attitudes about experience and education factor into it as well. She cites her last trip offshore when one of the crew members was a middle-aged man who was working on his PhD. It was only his second trip offshore and he was questioning how an aspect of the job was being done. Scarlett told him she was not the right person to ask and explained what the proper procedure was and reiterated that it needed to be followed. He became angry and responded by telling her he had been in the discipline longer than she had been alive and emphasized the difference in their degrees, "basically saying I've only got a BSc and he's got a PhD. Obviously I wasn't happy with this remark. I have always been taught to respect my elders, but that doesn't give him the excuse to get away with speaking to me like that." Saying it is the only time she has "really sort of had it out with someone offshore," she also kept her cool but voiced her opinion.

Upon her return home in March 2015, the industry was feeling the full effects of the downturn and, having purchased a home with the money she was earning offshore, she made the wise decision to accept a role onshore.

"I had just gotten a mortgage. I would have taken anything even if it was completely out of my field. But it's just an ethic, again, isn't it? I guess if you don't have that, you'll think, 'Oh, no, I'm only taking one that pays the same, and is offshore.' Well, you can't be like that at the moment.

"You've got to be proactive and you've got to think ahead. When you don't is when you get into trouble. Some [people] are just so blasé about it," she says, citing the case of a former crew member who was made redundant and hadn't told his wife. "He was going to tell her when he got another job!" When Scarlett spoke to him months later, he still had not found one.

While Scarlett was offshore in Mozambique, the Offshore Blondie blog was garnering attention within the industry and she was assigned articles for *Rigzone* and *Offshore Energy Today*. "One of the guys was like, 'It's alright for you; you're a girl,' and I said, 'No, it's because I've actually put in the effort to start something and stick at it. I don't *have* to be doing this.' That's the thing that they miss. They think, 'Oh, it's great that this has all happened to you.' You put in the effort and it can happen to anyone."

Her blog has opened other doors within the sector. Her writing has appeared in prestigious industry publications and landed her a regular column in *The Prospector*. She was asked by the UK Government to participate in its Not Just For Boys campaign, which she says, "is all about choice, inspiring women, and dispelling the myth that some industries are for men-only." She was invited as the guest of honor to open the Offshore Energy Exhibition & Conference (OEEC) 2015 in Amsterdam, assisted with the official opening ceremony, and gave her "first ever" public speech during which she discussed women's roles in oil and gas. Obviously, the organizers were eager to have the perspective of a millennial – the new offshore generation – especially one with knowledge and experience in the industry.

At the start of her career, Scarlett thought she might pursue a master's in soil mechanics and, in fact, was accepted into two programs, but once she started working offshore, "I just loved it. I got my break, and I've been really fortunate, and I am really proactive and passionate about it." So, for now, further formal education is on hold, although she says, "Never say never. Circumstances might change. I'm a woman; I might want children, say, in 10 years' time and take a break from offshore and go back and do that then. But, currently, I'm loving work and life."

The lure of offshore is difficult for her to resist, but she was able to look at the big picture when she was offered a short-term assignment in Canada this past summer. She says it broke her heart to turn it down, but she told the company, "You're offering me six weeks of work, fantastic, the money's great, *but* what's the rest of the year? It's still so uncertain."

For now, she remains onshore, her grandfather's words "work or want" always in her mind, although she is candid it is not where she would prefer to be. "I'm still doing geology, which is great, so I am doing something I'm genuinely interested in. But it's not for me. Like some people can't go offshore. The thought of going offshore is hell for them. They think I'm completely mad for doing six weeks. I would jump at it. I just love it. I'm *itching* to go offshore and I will do as soon as I can do," she says with a sigh.

Once she is back on the water, Scarlett has goals she intends to pursue. When she joined Benthic at the age of 22, the possibility of party chief training was mentioned. "I love geoscience, I love what I'm doing, but I'm very organized and do want to go down the managerial route and obviously do it offshore, so I do want to be a party chief. But, at the moment, I've said to them, 'I'm taking a step back for a couple of years.' *It's not the female thing; it's the age."*

Despite the image the title conjures, a party chief is actually a project management role "interacting with people and clients, and being a female, the clients always want to have a conversation with you, so it puts you that little step ahead. I just think enough comments have been said in that respect that sort of dampened how I feel. I think at 24, you're going to have a couple of people that won't take you seriously enough, even though I'm more than capable of doing that role. It's sort of put me off a little bit at the moment, which is a shame. I've still got plenty more years, I guess," she muses before reverting to her usual optimism.

"It's a gut feeling and I'm a true believer everything happens for a reason and your path is mapped out. I know it will happen; it's just a case of when, really."

♀ ♀ ♀

By the summer of 2016, Scarlett was back on the water, where her work has taken her to Senegal, Qatar, the US Virgin Islands, Ghana, Mozambique, and, most recently, Western Australia, where she planned to return in early 2019 for follow-on work as the lead geotechnical engineer.

<table>
<tr><td>

19

</td><td>

Jennifer DiGeso Norwood
First Female Offshore Installation Manager,
Maersk Drilling

</td></tr>
</table>

"Women still undervalue themselves; they don't
completely recognize the contributions they can make."

❝ I grew up with three brothers and it taught me that I can do anything and be anything. That was the gift they gave me," says
Jennifer DiGeso Norwood. "I never looked at anything and
thought, 'I can't do that because I'm a girl.' I always thought, 'Anything's open to me.'

"I had a mother whose attitude was, if they can do it, you can do it.
She never said, 'You can't do that because you're a girl' – never. Growing up I was ignorant to the fact that there was a mindset like that."

Born and raised on the Jersey shore on the US east coast, Jennifer
developed a love of water early. By high school, her family was living in
North Carolina – "a little bit inland" – and she decided no matter what
she studied in college, "I wanted to be by the water."

When she was invited to tour Texas A & M Maritime Academy in
Galveston on the Texas Gulf Coast – one of only seven maritime
academies in the United States[70] – she says, "I had no real knowledge of
maritime schools, but when I got there I fell in love with the ships.
Initially, I went in as a marine science major with the license option, but
decided to transfer to the marine transportation program because I
wanted to be onboard ships and in the shipping industry more than
discovering the science of the marine environment."

---

[70] The other six are: California Maritime, Great Lakes Academy, Maine Maritime,
Massachusetts Maritime, SUNY Maritime, and the US Merchant Marine Academy.
Source: US Dept. of Transportation – Maritime Administration

Starting with an intake of 30 women – the highest number up to that point –Jennifer was one of about 10 to complete the program.

Sadly, her father, who was always supportive of her pursuits and would remind her siblings of the sacrifices Jennifer made by sailing during Christmas holidays instead of going home, passed away during her senior year of college and she was forced to delay getting her license by about six months, making her official graduation date May of 1998.

Upon graduating with a degree in marine transportation and a third mate unlimited US Coast Guard license, Jennifer initially began working shore side in imports and exports, but after a year the thought occurred to her, "If I'm going to go to sea, now's the time to do it," only to discover there weren't a lot of ships available in the late '90s, resulting in her first ship being a union job.

"I absolutely hated the union," she recalls, citing the more formal nature of shipping compared to offshore, and having to deal with both licensed and unlicensed crewmembers.

There was also a lack of facilities for women at that time because so few women worked offshore. "The first vessel I was on had one public bathroom that was up by the bridge and that was the only place I could go to the bathroom. There was no women's changing room. I bunked with the guy who was the cook and it was kind of awkward to be in the room at the same time." Jennifer says the current generation "has no clue how comfortable their lives are now compared to what it was like before."

Disillusioned after sailing for three months doing classified government research in the North Sea, and upon hearing that the oilfield was looking for licenses, she decided to jump ship, so to speak.

She was hired by Global Marine in 1999 and says when she took her first test to go offshore it was at a facility where the employees actually had to do physical tasks. Initially she was assigned to go to a drill ship and, although the company wanted to hire her, it re-evaluated because at the time the physical requirement was carrying 89 pounds – "not just once but multiple times in a warehouse in a very hot humid environment going up and down stairs." Her tasks were stopped when the trainers felt like her heart rate was too high.

"I said, 'That's normal for me,' noting that she is 5' 1" and weighed 115 pounds at the time.

"The actual test was not representative of what happens offshore and is no longer realistic. You're not lifting tremendous amounts of weight anymore. You're being encouraged to think smarter not work harder. Being a woman, having worked on the deck, I didn't lift lots of things; I used a crane or rolled it down the deck. You find another way to do it. We don't expect anybody offshore – man or woman – to carry 89 pounds. There are cranes for that." Jennifer thinks the misperception that offshore workers need to be "big, strong, and tough" may make the sector less appealing to women.

"When I started with Global Marine, we had our own catering staff so we used to be able to take the mainly female employees from catering as the entry-level onboard the vessel into roustabout, so there was an avenue for advancement." She notes that most companies have moved to contract catering so that route is no longer available and roustabout has become the entry-level position.

Perhaps leading to further confusion about offshore opportunities for women, she finds people are often uncertain about oilfield terminology. "They have no clue what a roustabout is. When I went offshore a roustabout was a "roughneck." Nowadays they're called floor hands. They're not muscling around the equipment that much anymore. You teach people to be smarter – and safer."

Safety is an area where Jennifer believes women may have an advantage over men. "You can see outside of the problem because you're working from a different mindset to solve the problem without it being hazardous or having safety issues. You find ways to adapt – "think laterally" may be a better way to say it – and see the opportunities, and that comes from being a woman who's not muscling things around."

It was an important mindset to have when she hired on with Global Marine as a ballast control operator in charge of keeping the balance and stability of the vessel through the use of a "ballast," a compartment that holds heavy material (in this case water), and began working her way up, increasing her rank by sitting for the next US Coast Guard test (2nd Mate Unlimited), and "being in the right place at the right time."

Having grown up being treated as an equal to her three brothers, Jennifer was shocked to discover there were men who thought she was in the *wrong* place. "It's difficult being a woman in the industry; it's still very much a man's world. We still have a good ole boys' network. It comes down to the strength of the woman. You've got to be willing to say, 'I'm here to make a living just like you are.'"

Having found management to be "very open and supportive" when she worked onshore, Jennifer was surprised to encounter a different atmosphere offshore, some of which she attributes to varying levels of education. "It's a mixed bag out there and you're seeing guys who are in really high positions who don't have the experience and background. In this industry, as a woman, you've got to work twice as hard to prove you can do it."

In doing so, she very quickly mastered being a ballast control operator and, tired of the repetitive nature of the job, was ready to continue moving upward. "I'm always learning and looking for the next thing."

It was then that she says she faced the only challenge she experienced early in her career: getting passed over several times for promotion. "I really had to force my own hand by being in position, getting my license – or close to it – and going through management and eventually human resources (HR) and asking if they had other opportunities."

Her goal was to get her master's license, although she says it wasn't so much to become a master as it was to get the license, and she refused to sit around waiting while her career was dictated by someone else's time table.

Her advice? "If you're not getting promoted through the proper channels, you've got to figure out what the next avenue is and, if you're a valued employee, then you should get that opportunity."

In Jennifer's opinion, the oil and gas industry – offshore particularly – was not very proficient at establishing the promotion rate at that time.

"It used to be whoever was somebody's buddy," she says bluntly. "And then as everything started to really boom, the demand for people grew, but so did the need and the knowledge base, and I was in the right place at the right time in my career. Whereas now you're seeing a lot more women offshore, there were times when I was the only woman and then there were times the only other women offshore were the ones in the catering crew."

That dearth of women also meant her mentors were all men. When she was second mate, she found herself working under a Canadian man named Andy Allen, who had been a captain, but stepped back to be chief mate on a drillship because as she explains, "Not everybody had drillship experience and it was a big thing.

"He definitely had the whole captain persona," but in illustrating one of the best case mentor-mentee scenarios, Jennifer says, "I got to learn from him and he got to learn from me."

There were two other male mentors who would be influential later in her career. Scotsman Brian Morris and Ian MacKay from Scotland – captains on the same ship – gave her opportunities when she says quite simply, "They didn't have to.

"Andy was going to be our captain and I was going to take his place as chief mate, so he was influential in opening that door for me to prove I could do the job in order to move up. When I did and earned my way then it was an easy decision for Brian and Ian to give me the opportunity."

In 2003, Jennifer was devastated to learn that Andy had perished in a diving accident. "I'll never forget it – we were offshore, coming back from Australia on the ship GSF[71] *Jack Ryan*, steaming across, and I saw a newspaper article that had been printed from the Internet saying he had passed away. I couldn't believe it.

"It's so important to listen to people," Jennifer stresses. "He sat down with me – I had come out of academy and was gung-ho and had high expectations for young third mates coming out of school who were failing in AB[72] positions – and said, 'You're a little hard on them.'

"I said, 'I am not expecting them to do anything I wouldn't do.' And he said, 'Yeah, but your approach is too hard.'" It was a moment of revelation for Jennifer who says, "It was the first time that I actually had to evaluate myself. And that began my start in reading leadership books."

While she doesn't recall the title of the first book she read, she remembers it was about "being the boss." From there she went on to read how to develop the leader within and then how to develop her team to become leaders.

"If I could impress upon one thing – because unless you take courses in leadership, schools don't teach it – it's how do you learn it, where do you learn it? Especially early in this industry, most of the people who moved up didn't move up because they were good leaders, they moved up because of brute force. It used to be, 'If you don't do the job, I'll run you off.' We were expecting them to manage people and they didn't really know how without being bears and bullies."

---

[71] GlobalSantaFe Corporation (GSF) is the world's second largest offshore drilling contractor. Source: GlobalSantaFe Corp

[72] Able-Bodied (AB) seaman, the entry-level deck position aboard any vessel.

By the same token, she didn't realize the quiet confidence she possessed could be just as intimidating. "Looking back now as an adult I probably was intimidating to people because I had goals and a path forward and I knew where I was headed. A lot of people don't have those things figured out." As Andy Allen once assured her, "You know who you are and what you're doing."

When coaching, mentoring or training people, Jennifer says she always asked herself, "What kind of officer do I want them to be?"

The advice she gave offshore was: "Know your position, learn the position above you, and teach the guy or gal behind you. That was always a tough one. People don't want to teach somebody; they don't want somebody to take their job. I never felt that way. Any information I can teach you, I'm going to give you. In some respects, this industry hasn't changed but, in terms of new technologies, it really has. You want to be ready for that next position when the opportunity comes and if you haven't taught somebody behind you to take your slot, you're not going anywhere."

In 2004, at the age of 30, Jennifer fulfilled her dream of becoming a master. "One other gentleman who opened the door was my first rig manager Fred Brooks," who in 2017 became executive vice president, operations and engineering with Rowan Companies. She credits him with "really making it happen when it came time to be an offshore installation manager (OIM)."

Her first OIM job was on the GSF *C.R. Luigs* with GlobalSantaFe. "I was the first female master/OIM for GlobalSantaFe and was onboard when BHP did the story and discovered I was the only female at the time working as OIM anywhere in the world. There had been one other female before, but she only lasted two trips.

"As OIM and as the captain, you have the final say." And you cannot be afraid to use that authority Jennifer says. "In my position as a captain I've had to pushback on clients and they didn't like it. You need to know where your line in the sand is, what you're willing to push back on, what you want to fight for." As she points out, "When you're the master, you potentially are held liable for the safety of the vessel at any time. It's my head, my license, and not just my company's, although that's who I'm representing."

In 2007, after eight years with GlobalSantaFe, she left during the merger with Transocean and moved onshore as the Quality Management, Health, Safety, and Environment (QMHSE) Manager and Director for Maersk Drilling. There it was her job to "influence how people are thinking from a safety standpoint and open their perspective because, if you can look at things from a different perspective, you can see where the hazards are. If you have a closed mind, you're never going to change your behavior.

"People say, 'accidents happen,' but they happen because we're human and because of a series of things that cause them to happen. It's a chain of events and you learn from it. When you don't learn from it, there's a problem."

A year later, Jennifer became Maersk's first female OIM and also the first US citizen to be granted a Danish license to sail a Danish-flagged vessel, as Maersk's masters and OIMs were required to be EU citizens. She was often met with disbelief by crew members who would say, "I never thought having a woman OIM was going to work for me, but you have surprised me; I'm amazed."

And again she says, "You earn their respect." Coaching and leadership were two aspects of her role offshore that she enjoyed the most. "There's a fine line in how you manage people and lead them because the reality is you want them to do something that needs to get done, but you don't want them fighting you tooth and nail. The reward is they're going to win, they're going to have their own responsibilities, and you don't need to chase them all the time." As a leader, she says, "The biggest thing is learning to talk to people, listen to them, and hear what they have to say."

Her last job offshore was as a master with Pacific Drilling Co. and, by that time, stress was taking its toll. "The reward versus the stress wasn't there." While Jennifer says imparting knowledge, coaching and guiding the people under her charge were the rewarding aspects of the job, the greatest and most difficult challenge was the generational change – this from a woman who was only in her 30s at the time.

"The younger generation was coming up and it was already hard enough to be in a position of authority as a female and have worked and proved your way. Some of these younger men, who move so quickly into positions that maybe they're not ready for or who don't have a full understanding of their position, think you didn't earn your way, so it was a strange transition in that sense. The generations that are coming in now kind of have the attitude, 'What are you doing for me?' Stress is high and people come to you with lots of problems and no solutions.

"The fun part and the challenging part of what you're actually doing get really small and you're not learning anything."

An element of the conflict was Jennifer's personality and another was the nature of the business. "Just before I met my husband Adam, I was looking to go into rig management and move up that direction, and then I met him and I thought, 'What do I want to do long term?'

"Being the person I am, I almost saw myself as having to choose between having a family or a career. Before I met my husband, a woman in my position, making the salary I was making, was very intimidating to a lot of men. It's difficult, too, if you're going to have a family because you have a limited number of years that you're going to stay offshore.

"When I was working, I always thought, 'What is the top level going to be for me? When am I going to be satisfied with where I am?' And I don't know if I ever would have been."

After a 12-year career that took her to the Gulf of Mexico, Trinidad, West Africa, Australia, and included some time in the Samsung Heavy Industries shipyard in Geoje, South Korea, Jennifer retired in 2012 at the age of 38.

"Actually it probably worked out for the good; it was time for me to move on."

Once she retired, she was sought out by a woman whose Health, Safety, and Environment (HSE) company used to work on the rigs Jennifer oversaw. Over the next four years, she began doing technical writing for the woman's company and working on a book about her career. Being self-employed and working from home enabled her to take care of her mother, who was terminally ill, as well as spend more time with her three stepchildren.

"Of all the jobs I've ever had, being a stepmother is the hardest," she says frankly. "I have a sister who's 10 years older than I am. She says, 'You always had that confidence about yourself, you just always knew.' It's very foreign to me to think that people don't have that. And now that I have three stepchildren – including a 10-year old daughter – I get it."

Despite her offshore career and her accomplishments in a male-dominated field, Jennifer says, "I can see that my daughter thinks you should only do certain things as a woman. Especially in this day and age, it's so hard with the way women are portrayed, the sensationalism of Hollywood and reality TV, and so many other influences like social media. How do we make math and science compete with that?

"I think that women in a sense still undervalue themselves. They don't completely recognize the contributions they can make" – something she sees as a societal problem. "As I get older, I can see that attitude is still very much there. Women in other industries that aren't even male-dominated still deal with that kind of mindset because, as a society, we have thought men were the breadwinners and women stayed home and were the caretakers. And even though the reality has changed, the mindset hasn't."

In 2016, she joined Maine Maritime Academy as a lab technician and teaching assistant. After about a year and a half, she became an instructor while simultaneously working on earning her Master of Science, International Logistics Management. Her quest for that next level hasn't ended and she plans to pursue a PhD while continuing to teach. "In my pursuit, I'm always looking for ways to improve the education of the students."

Reflecting on the gender gap in the industry, Jennifer says, "I always think it's two-fold. Part of me feels like in one respect you're personally filling a gap. And why is there a gap? The reality is, most people say, 'You wouldn't be talking about it, if she wasn't a woman,' but we wouldn't have to talk about it if it women had been more accepted in every industry and not considered the weaker sex. Raising awareness is not a bad thing, but I think letting women know that they can do anything and be anything is really the goal."

♀ ♀ ♀

20	**Sue Jane Taylor** **Fine Artist & Visual Documenter**

"It's a very different world out there,
which we know nothing about onshore."

❝ Watching these sailors moving about like acrobats, loading and unloading these huge containers on and off the ship, whilst the boat was swirling up and down, round and round, and the waves coming onto the deck." It is not surprising that, as an artist, Sue Jane Taylor would speak in such poetic terms about her first sight of an offshore oil platform more than 30 years ago.

What is surprising is the stark contrast to the view that young women still held in 2015, according to a Price Waterhouse Coopers' (PwC; 2015) survey in which 57% of female millennials stated they would avoid working in a business sector that had a negative image, including the oil and gas industry.

Sue Jane herself didn't envision the offshore world as a place of inspiration and beauty in 1984 when she was invited by Stirling Shipping Company in Glasgow, Scotland, to visit an offshore installation as one of 12 artists to commemorate the 10-year anniversary of the company's involvement in the North Sea. She recalls boarding the *Stirling Teal*, a cargo vessel, on a very dark, stormy December night in Aberdeen.

"That voyage, that journey, that week, was just an incredible visual stimulant for me. Not only being in an all-male environment with tough sailors and trying to get to know the guys, but out in nature – seven, eight, gale-force weather – and being tossed and turned. It's a very different world out there, which we know nothing about onshore. That was my first impression, my first experience being offshore and really, I got a taste for it."

Photo Credit: John Paterson, Murchison's Helicopter Landing Officer

In what by now has become a familiar refrain when talking about women in the oil and gas industry, Dr. James Hunter[73] writes, "As Taylor herself tells it, her long engagement with oil-related industry... began almost by accident" (2013).

While she may not have thought consciously about focusing her work on the petroleum industry, the images were the backdrop to her child-hood in the Black Isle in the northeast Highlands of Scotland, where she was born and brought up. She recalls it being "a quiet sort of backwater in many ways," until the '70s when the oil boom "radically changed everything: socially, geographically, economically – and it was all around me."

She remembers the Black Isle being flanked by construction yards, each with several thousand men working frantically to build the fixed steel jackets, top modules for the North Sea oilfields. More fabrications yards had been set up to the west in Loch Kishorn, Wester Ross, and Arnish, Isle of Lewis.

"The whole area was in flux and as a child I experienced that. Then when I attended art college in my undergraduate, I was in Aberdeen, the oil capital of Europe, and I was affected by the oil there in the sense of very expensive accommodation, hard-to-get digs, and also the presence of oil guys everywhere – in the bars, night clubs, discos – you couldn't get away from it."

Although Sue Jane was surrounded by the oil and gas industry, it still felt remote to her because, up to that point, her work had not been influ-

---

[73] Emeritus Professor of History at University of the Highlands and Islands, Inverness, Scotland

enced by it; instead, her art had always related to landscape, specifically a northern landscape. The invitation from Stirling Shipping Co. would change not only her art, but also her life.

Her first trip offshore to a platform was a day trip in 1986 to British Petroleum's Forties field. "It was a whirlwind of a day with VIP treatment. I was shown all over the platform in a bustle of activity, meeting people at all different levels, and then returning to Aberdeen in the evening. I didn't have time to stop and contemplate, or even draw in my sketchbook. If I wanted to really experience and get the feel and atmosphere of the place, it was necessary to stay for a longer period."

At the time, Sue Jane was living and working in London, where most of the oil and gas companies were located. She had visited the oil terminal on the island of Flotta in the Orkney Isles, which was the terminal for the Piper and Tartan fields. Having established a good relationship with the PR manager in London, she approached him and asked if it would be possible to get onto one of the platforms in the North Sea in order to gain a more complete understanding of what it was like to be offshore.

"A day visit was no use," she says. Having managed to convince him, she spent a week on Piper Alpha, almost a year prior to the July 6, 1988, tragedy, which claimed the lives of 167 men (61 men survived; there were no women onboard).

"It was the same shift of men, who were on that night, during the disaster as well. It was an incredible week for me. It's still very vivid," she says nearly 30 years later.

"At that time, it was an all-male environment anyway. Compared to the Norwegian sector, there were hardly any women who worked offshore. The medic, Gareth Watkinson, took me to lunch in the large canteen. The only other woman on my flight out joined us. She was on a day visit. She was in charge of statistics for safety standards and this was her first time offshore. This platform had recently managed to achieve 100 days without any accidents. She handed out a pewter tankard with 'We did 100 days, Piper A Oxy, 28.05.87' inscribed on it. I work closely with the Aberdeen Maritime Museum and I thought it would be good to give it to a public institution, so I donated that tankard to the museum.

"When I was on Piper, I was given the freedom to get outside unaccompanied whenever I wanted and it was quite extraordinary. I mean, that would never happen now; it doesn't happen now. You always have a "minder" with you, accompanying you everywhere."

"I rested until 10:30 pm, then went outside to walk around and take some infrared shots on the deserted platform levels. It was a dry but windy night, and my only companions were the noises of the platform machines as I walked up onto the deserted helideck. Looking around from this vantage point, I could make out the neon lights of the sister platforms belonging to Occidental, Tartan and Claymore. The lights on the deck illuminated the words 'Piper A,' and looking out into the blackness, I could see the delicate flickering lights of the safety vessel on the pitch-black sea. I angled my camera down into the deep, dark pit, which was the sea. The noise of the waves rushing against the well conductors. It was like a strange, timeless dream, with the acid lights shining onto the metal against the forces of the sea. No one could hear my cries if I fell overboard."[74]

–SJT diary entry dated July/August 1987

Two years after the Piper Alpha tragedy, Sue Jane would be asked by representatives of the families and the Piper Alpha Families and Survivors' Association, as well as the Aberdeen City Council, to create a memorial to honor their lost loved ones.

"It was a great honor and very humbling. Deep down, I felt like I was the right person to work on it. It was really hard because throughout I tried to focus on my art, on the actual piece, and keep out of the politics."

Using live models, including Bill Barron, a Piper Alpha survivor, whose likeness inspired the "mature" central bronze figure, she began with drawings in conté and charcoal, from which she would eventually sculpt the three figures in plaster.[75] Nine months later upon completion, they were cast in bronze by the Burleighfield Foundary in Buckinghamshire. On July 6, 1991, the memorial was unveiled by the Queen Mother at a ceremony in the North Sea Memorial Rose Garden at Hazelhead Park in Aberdeen attended by 1,000 people (McGinty, 2008).

When it came time for the unveiling, Sue Jane was overwhelmed with emotion. "It was a beautiful day in terms of the weather, the sun was shining. The site is a beautiful place, very calming in a space of its own

---

[74] Excerpted from *Oilwork North Sea Diaries 1984 – 2004* (Birlinn Publication; 2005). Reprinted with permission from the author. All rights reserved. © Sue Jane Taylor.

[75] The other two men who volunteered to be live models were a sculpture student (roustabout figure) and an arts graduate (survival suit figure).

with a rose garden surrounding it. Every rose bush was chosen by the families for its color, its scent, and its name in coordination with the head gardener. It's a universal homage for a lot of people, not just for Piper Alpha victims, but for all families who have lost someone in the North Sea."

Sue Jane would return to the Piper field in 2004 at the invitation of Jacquelynn Craw, then legal manager of Talisman Energy,[76] who commissioned her as part of an oil work exhibition, believing it would be an invaluable historical record for an artist to document the experience of being on the new Piper B platform.

"I was very reluctant at first, but I trusted her gut feeling about it. It was a really powerful week for me and also a reconciliation in every way. It was good to be on a platform there and to speak to the guys and also some of the women who worked on that shift at the time."

"We walked around to the north side of the platform, on the drilling deck side, and [the offshore health & safety advisor] pointed out the Piper Alpha buoy marker. I was shocked how close we were to the original site of Piper Alpha platform – only 600 metres away. Nothing remains of it. Most of the sub-sea leg structure was taken away. I was numb in every sense, and all the pain of the aftermath flooded back. My thoughts were drawn to the families and survivors I had got to know. How lonely and desolate the marker looked, bobbing and flaying on the surface of the rough, grey sea. The reality of that horror struck me full in the face."[77]

–SJT diary entry dated September 2004

The following year, Sue Jane would publish a compilation of her diary entries in *Oilwork North Sea Diaries 1984 – 2004* (Birlinn Publication, 2005) and says, "My diaries play a vital part of my work. Oil companies like tight corporate control and there has been so little objective visual documentation and observance. I've brought another kind of visual perspective and viewpoint, which I think compliments the corporate view.

---

[76] Acquisition of Talisman Energy, Inc. was finalized by Spanish integrated energy company Repsol on May 8, 2015. Source: Repsol

[77] Excerpted from *Oilwork North Sea Diaries 1984 – 2004* (Birlinn Publication; 2005). Reprinted with permission from the author. All rights reserved. © Sue Jane Taylor.

"I've been very fortunate throughout the years to get avenues and opportunities to go into these offshore installations and places, where the public are not allowed. It is a privilege, which I find absolutely..." she trails off and then, despite the eloquence of her diary entries, says, "Words are difficult for me because I'm a visual artist, but I try to disclose it in my work."

Her fascination with life offshore and the men and women who inhabit that world has never waned. "Being on a platform is so inspirational; it's the power of nature. It's people working in these extreme and remote environments and how they react to them. It's a psychologist's dream; it's a fascinating observance of human nature.

"Artists have always had this fascination for the symbolic, the concept of the source. There's the source of the black gold, and it's out there in these remote platform installations," she says, creating an aura of mystique and exoticism not often associated with the petroleum industry.

In 2006, she was contacted once again by Jacquelynn Craw, this time for a different type of offshore commission. Talisman Energy had become involved in the Beatrice Project, an offshore wind farm proto-type project, in the Beatrice oilfield. At its nearest point, approximately 13.5 kilometers off the coast in the north of Scotland, it's the oilfield closest to shore in the UK sector, and can be seen from land.[78]

Calling it a "truly European project," Sue Jane traveled to Denmark and Germany, where the components were being built and assembled, and observed the men and women at work manufacturing the blades in Lunderskov, Denmark, and also Husum, a maritime town in northeast Germany, where the generator for the 5M turbine – at that time, the biggest in the world – was being manufactured.

"It was just fascinating being involved in that project because it was basically oil technology being transferred and used for renewable energy."

Beatrice Works, her drawings, sketches, and portraits from that time, were shown as a touring exhibition from 2013 – 2014. At one of the venues, the Caithness Horizons Museum, the education and commun-ities officer created a program encompassing different aspects of the local industries, which were involved in the nuclear, renewable, and oil sectors. A key part of the program was for the students to go on site

---

[78] Source: Beatrice Offshore Windfarm, Ltd.

visits, where they were treated as though they were a customer or contractor.

They met engineers, toured a deep-sea diving vessel, and spoke with the divers who were still in their decompression chambers. "It was a bit of an eye-opener for them, especially the secondary school pupils," Sue Jane says. She also observed that when the representatives from the renewable energy industries gave their presentations, some of the girls expressed an interest in the engineering side of it.

While there is a great deal of emphasis on STEM these days, Sue Jane believes art is a crucial component, creating the acronym STEAM. As a participant in the museum's program, her role was to help the students experience being an artist and to learn how to record and document. "I hope it made some impression on the students to know that arts can be part of science and technology, and have an important role, and also to meet an artist who does something different from what they think an artist to be.

"For every touring exhibition or residency, I try to involve some aspect of the education program, whether it's with adults or students or young people in further education. I think it's important to see how an individual artist works."

Sue Jane's work has never focused solely on the offshore realm. Even though she continued to work on the Beatrice Project through 2012, she began a two-month residency in 2008 as an international visiting artist at Landscape Artist Research Queenstown (LARQ) Tasmania. There she spent time underground documenting the lives of men and women in mining, an extractives industry that employs fewer women than oil and gas, although she discovered quite a few females – Australian migrant mine workers called "seagulls" – are hired for their driving skills ("they tend not to crash and bash the vehicles against the tunnel walls"). In stark contrast to her work offshore, where she is surrounded by the elements, she recalls, "I had drawn many times in industrial environments, but never one so extreme, so devoid of light (2009). You can feel the psychological weight of the rock overhead."

Her work began in the relatively early days of the UK offshore industry. It would come full circle in 2014, a decade after she had last been offshore, with the decommissioning of some of those same installations. She credits Carol Barbone, who at that time was the decommissioning lead for Canadian Natural Resources International (CNRI), owner/operator of the Murchison field, for helping make that possible. Like Jacquelynn Craw, she saw the urgency in Sue Jane's visit offshore, which enabled her to witness Murchison's last months of production, cessation of production, and be part of the last crew of 17 before the platform's final abandonment.[79]

"It was wonderful to get back onto a platform again after ten years' gap. It was inspirational to me. I was there to witness the cessation of production after 34 years. I met some of the men who had been there for the whole of the 34 years, and it was emotionally charged when I was talking with them and trying to record their working lives. There was a great sadness, almost bereavement, in this platform being decommissioned, broken up. It was again a privilege just to be there and to meet these people, and to try and record some aspect of production before it was shut down, and then on the second trip to witness the start of the decommissioning."

While Sue Jane documents the lives of offshore workers, regardless of gender, she does make it a point to talk to the few women she comes across on the installations. "I am drawn to the women who work there because there are not many, so you ask, 'Why are you working here?' and, 'How did you get into it?' because it is a very unusual occupation really.

"It's trickier for women if they have children. But a lot of women whom I've met offshore have had children, and it's very complex, but they kind of work round that. They have to work that extra bit harder, as we always have to do, in order for it to work, don't we?

"When I recently was speaking at the British Science Festival in Aberdeen, I said, 'I think there's a lot more women working offshore.' This woman in the audience corrected me, saying that only three percent of the North Sea work force is women.[80] That's quite telling."

For Sue Jane, the offshore world holds a special allure, something that transcends gender or stereotypes, the mysteries of which she strives to share with the outside world through her life's work.

"Being on the platform is a timeless, surreal sensation. It could be a futuristic space age or it could be any time in history because there's nothing... no time... nothing to gauge, just the horizon, the sea, and the sunrise and sunset."

---

[79] Decommissioning activities began in 2013 with the plugging of the wells. Production ceased in 2014. The platform was removed in 2017. Source: Norwegian Petroleum Directorate's FactPages

[80] According to the *UKCS Offshore Workforce Demographic Report 2015*, women continued to represent 3.6% of the total offshore workforce as in the previous year. Source: Oil & Gas UK

Description of figures

The central bronze figure, which faces north towards the main entrance of the gardens, represents a mature character. In his left hand he holds a pool of oil sculpted in the shape of an unwinding natural spiral form. This black shape in his palm flows into gold leaf. His right hand points down to the ground, indicating the source of the crude oil. The carved motif on his helmet, a fish and seabird design, symbolises the environmental aspects of the oil industry's presence in the North Sea.

The 'roustabout' bronze figure, which faces west, represents the physical nature of many offshore trades. His pose emphasises two opposite 'strain' movements in offshore work: push and pull. On his right sleeve is a 'tree of life' motif, based on the Celtic design. Its leaves are in gold leaf. The design's mytho-logical meanings symbolise the exploration and production of crude oil. For example, its roots deep in the bowels of the earth represent the vast oil wells underneath the sea-bed; the tips of its branches reaching up to the sky, the eternal flame of the flare boom on oil and gas production platforms; the vapour rising into the earth's atmosphere.

The 'survival suit' bronze figure, which faces east, represents youth and eternal movement. On the left sleeve of this figure is a design of a sea eagle's wingspan and its head, gilt in gold leaf. The sea eagle is native to the northern seas, and it is used in place of the North American eagle, the patron of oil.

On the south face of the Memorial plinth, above the Celtic cross, are inscribed the names of the thirty men with no resting place on shore. A casket of unknown ashes is interred behind the cross. On the east face of the plinth are inscribed the names of the two heroic crewmen of the Sandhaven rescue vessel (2005).[81]

---

[81] Excerpted from *Oilwork North Sea Diaries 1984 – 2004* (Birlinn Publication; 2005). Reprinted with permission from the author. All rights reserved. © Sue Jane Taylor.

| **21** | **Pat Thomson** |
| | **Materials and Logistics Superintendent** |

"Because of my age, I was almost an icon."

Nowhere is the oil and gas industry's gender imbalance more apparent than offshore. The annual *UK Continental Shelf (UKCS) Offshore Workforce Demographics Report* (Thom, 2014)[82] put the number of women at just 3.6% of the total offshore workforce and, of those, approximately 30% were employed in catering, not operations. In actual numbers, this means that of the 61,892 people employed offshore, only 2,237 were women. The highest percentage of women was found in the 24 – 29 age range, decreasing steadily from the 40 – 44 demographic onward until age 65+ when there were only three women employed offshore in the UKCS.

♀ ♀ ♀

Wearing a gas mask to prevent exposure to deadly hydrogen sulfide (H2S) gas.

Waiting for an emergency helicopter to evacuate the crew from a burning rig. Gale force winds of 90 mph and seas up to 17 feet. Not exactly the conditions under which dream jobs are made, but Pat Thomson, a materials and logistics superintendent living in Derbyshire, England, can't imagine doing anything else. In fact, she originally went offshore to escape working life onshore, which she felt was "getting a bit mundane."

Pat – one of the three women mentioned in the 65+ age group of the UKCS report – is now a 73-year old mother of three, grandmum of nine, and great-grandmum of two. After leaving school at the age of 16 "with

---

[82] The most recent year for which figures were available at the time of the interview.

no qualifications," she began her early career as a research assistant working in a laboratory under the direction of renowned, somewhat eccentric British scientist, Dr. Magnus Pyke, "trying to speed the maturing process of whisky to enable faster returns." During that time, she was persuaded to attend college and, in two years, earned a National Certificate in Chemistry and Physics.

Photo Credit: Jean Paris

She eventually left work to start a family and, after having three children in four years, attended night school to gain Scottish Highers in Mathematics and English.

Once her children were in school, she was admitted to a three-year, full-time course for primary teaching and, upon completion, immediately secured a teaching post.

After four years, and the realization that teaching the arts to small children wasn't her strong suit, she looked elsewhere for employment.

The oil industry was in its infancy in Peterhead, Scotland, where she was living at the time, so she began cold-calling personnel officers, hoping to be offered a position. "Nothing ventured, nothing gained" was her philosophy and, although she was prepared to go door-to-door to achieve her goal, it didn't come to that. She does admit, however, to using "Pat" and not her given name of Patricia on her CV, knowing, at that time, people would assume Pat Thomson was a man.

She was hired by Brown & Root for a staff position and worked on a project building the now-legendary Forties Field on behalf of British Petroleum (BP), during which time there were regular visits from BP management to monitor the progress of the project. Pat's abilities and work ethic did not go unnoticed and, as the project was completed and the platforms handed over to BP, she was offered a post directly with BP, hiring on as a full-time employee.

After a while, she discovered she wasn't being promoted – not because she was a woman, but because she didn't have the necessary educational background – and realized she needed to earn the proper qualifications. Through correspondence studies, she earned a Higher National Certificate in Business Studies and later studied foundation, pure, and applied maths, physics, and engineering materials through

Open University in the UK and, upon completion, was awarded a Bachelor of Science.

It took years of study, but the investment paid off.

Within a short time, Pat was promoted from expeditor to engineering buyer and then senior engineering buyer.

"I used to get invited to all the functions. The reps wanted to take me out to lunch because, at the time, I think I was the only female engineering buyer in Aberdeen. The other buyers, when you managed to reach the purchasing departments of the big oil companies, were stationery buyers. That says it all, doesn't it? The drilling or electrical buyers were all male. It's come a long way."

She later became the onshore coordinator responsible for the logistics of three jack-up rigs in the southern North Sea with occasional trips offshore for end-of-well inventories. Pat is of the firm belief that even those employed onshore benefit from going offshore in order to better understand their job and how it relates to what is happening on the rig.

After 11 years as a BP employee, Pat applied for redundancy. During the notice period, she was approached by a recruitment agency on behalf of Phillips Petroleum and offered the post of materials engineer, basically making her the right hand man to the drilling supervisor. At the age of 49, it was the beginning of her career as a consultant.

When she was first offered a job working an offshore rotation schedule, she wasn't interested. "Then they told me the day rate and I said, 'I'll try it!'"

The mandatory survival training, which includes being submerged in a pool of water in a simulated helicopter crash, proved to be a challenge for Pat, who has a paralyzing fear of getting water in her mouth. Although she decided, "I'm not going to let this beat me," her phobia was so intense, she enlisted one-on-one training with a lifeguard at a local pool.

Offshore workers are required to go through survival training every four years and each time Pat says, "Never again," and then finds the courage to do it. "The last time was in 2012. Now my family just laughs at me when I say I'm not going to do it."

While her education enabled her to get the desired promotions, Pat quickly discovered, "There are things you don't learn from a textbook." As the only woman on the rigs in the early days, she had no mentors – male or female. "The men were too busy with their own jobs. I had to teach myself and I learned by trial and error. I was terrified; one mistake and I could have shut down a rig."

She arrived at her first full-time offshore job on a Friday and freely admits, if there had been a helicopter back to shore that weekend, she would have taken it. Unlike her previous forays offshore, she had never been so far out in the ocean that she couldn't see land or lights, if only from another rig. Talking herself through an overwhelming sense of isolation the first few days, she has never looked back.

That's not to say there wasn't a learning curve.

"I was quite naïve," she says. "I hadn't been around roughy, toughy oil drillers." Nor did she know the lingo. She recalls the first time she was told to order "nipples," she was too embarrassed to say the word over the phone. Another time, despite the fact that it was July, when the offshore drilling supervisor told her he needed her to expedite the delivery of a "Christmas tree," she asked him, in all sincerity, if he wanted fairy lights on it. She offers a bit of simple advice, "Learn the terminology."

As a materials and logistics superintendent, Pat's job required attention to detail and accuracy. As such, she was responsible for organizing the helicopters and vessels to and from the rig, as well as maintaining an inventory of the materials and equipment – "so that we know what we've got on the rig at any particular time; it's not like you can go to the corner store and get it" – arranging personnel; and keeping track of daily costs. An important aspect of the job included adhering to health and safety standards, ensuring all paperwork and proper handling procedures were being followed. In addition, Customs work took priority, necessitating accurate completion of paperwork for import to and export from the rig, as well as to and from the country.

"I never did deck work," Pat says, "but I got my hands dirty." Nostalgia sets in and she claims, "I was quite pampered, really."

Dutch filmmaker Rob Rombout's 1994 documentary, *Black Island*, opens with the following words written in white text superimposed against a black background:

> In the North Sea today
> next to sixty-thousand people
> are living offshore.
> On board the F.G. McClintock rig
> eighty men and one woman work around
> the clock in search of Black Gold.

That one woman was Pat Thomson.

The stark opening is fitting for a film, which portrays a distinctly *un*pampered work environment – one that is cold, wet, and dirty – and one in which an unnamed narrator says, "is probably the only place left, the only type of industry, where you actually have to work, eat, sleep, and live in the same place for 24-hours a day for three weeks at a time." A summary of the film on the director's website calls it "a micro society behind closed doors."

At one point in the film, a man is shown – presumably, he is the narrator at the time, although it's not completely clear – and the voice-over says, "This sort of rig – a drilling rig – is not *[unintelligible]* for women." The camera pans to Pat, wearing the requisite hard hat, dripping in spray from the sea, while she consults her clipboard, as the narrator continues, "The blokes come out here to work and that's what they do. A woman's work ends out here. You know, the jobs we have to do, is *[sic]* not suited for women and a lot of people feel it isn't their place, and they're just taking jobs from blokes out here." Adding impact to his words, the director has the cameraman keep his lens trained on Pat – the only woman working on the rig – while the narrator makes these assertions.

When asked if she knew the audio and video would be juxtaposed that way, making it seem as if the man's comments are directed toward her, Pat says, no, but that she doesn't recall having any particular feelings about it when she saw the final version of the film.

"I know at one point the crane operator said, 'There's no place for a woman offshore,' and the film crew asked, 'What about Pat?' And he said, 'She's different; she's just one of the boys,' and I thought, 'Is that a compliment or isn't it?'"

In response to one of the men saying she had a "paperwork job," Pat says calmly, "I still had to get off the helicopter; I still had to go on deck and check equipment; I still had to be on a rig that [potentially] could blow up at any time; and I still had to cope with the weather conditions."

Nevertheless, Pat is adamant that she has never been the target of gender discrimination or sexual harassment – or even ageism, for that matter. She speaks fondly of life on the rigs, including the *F.G. McClintock*, referring to the men she worked with as "the lads," and tells how they ended up calling her "Auntie Pat," as she eventually found herself in the role of "agony aunt," listening and giving advice when the men became lonely and homesick after weeks offshore.

She often mentions the companionship and team effort that the work engenders and says, "It offers challenges and friendships and respect – provided you earn it." That respect sometimes manifested itself as protectiveness. Pat, who is "only little," as she describes herself, at 5'3" and eight stone,[83] tells how the men would form a human chain, holding onto the back of her survival suit, and pass her from one man to another until she made it safely from the helicopter to the rig floor as the North Sea winds howled around them. Otherwise, "I could easily have been blown into the sea," she says matter-of-factly.

Despite not having the relative safety of a permanent job, she has been steadily employed since she started working an offshore rotation schedule in 1990.

"Even during the downturns, I was headhunted," she says. Her reputation preceded her, resulting in drilling supervisors specifically requesting her for jobs and oil companies asking agencies to track her down between assignments.

"That's why I managed to work for so long – there just isn't the expertise available. I was very conscientious. I absolutely loved it. I had fun as well as working hard."

While she seldom encountered other women, aside from caterers, on the rigs, Pat sees no reason why more women shouldn't work offshore "as long as they can deal with the conditions." Of course, what some may see as a negative, she views as a positive.

"You have to be independent and able to manage on your own. You have to be controlled and organized. The rotation schedule allows you to do whatever you want on your days off; you have quality time at home. What other job would you get six months' holiday a year? You have to decide what you want."

The job also enables offshore workers to travel all over the world and can have some very unexpected perks. While working in Namibia, waiting on the rig to arrive, the company, fearing the employees might get bored and leave, paid for them to go on a safari. "It was phenomenal – and that was work."

Pat's assignments also have included the Falkland Islands, Equatorial Guinea, Poland – where she worked on a rig in a shipyard in Gdansk – Holland, and Gabon.

---

[83] One stone equals fourteen pounds (eight stone = 112 lbs.).

She concedes it can be difficult for women to work in cultures where men are not used to taking orders from women. In Gabon, where she was in charge of the project, it was a different story.

As Pat explains, the average life expectancy for males is in their early 60s. "Because of my age, I was almost an icon," she says delightedly. "I baffled them completely. They treated me like a queen and called me Madame Patricia."

Then there is the money. As a materials and logistics superintendent, Pat has made as much as £700 (US $1100) a day; however, she is quick to point out that, as a consultant working a rotation schedule, there is no pay on "days off." While commanding that kind of day rate no doubt provided a measure of financial independence, she says convincingly, "I loved the work so much, the money didn't matter. I would have done it for half that."

Pat retired in 2013, but quickly discovered she missed life offshore. While she doesn't deny the negative aspects of the job – in addition to being dirty, arduous, and inherently dangerous, it has left her with arthritic fingers and minimal noise-induced hearing loss – for Pat, it has been the career of a lifetime and one she hopes to continue. As she nears her 74th birthday, and waits to hear from a recruiter, she has a simple wish: "I hope somebody hires me. I want to go back offshore.

"I'm dancing four days a week, but it's not the challenge I need. I love to be in action."

♀ ♀ ♀

22	**Nina Vorderwülbecke** **Helicopter Pilot**

"I'm very happy with the way things have turned out,
but you have to make sacrifices."

❝ Maybe I never learned to fly the perfect traffic pattern," muses Nina Vorderwülbecke, referring to the series of maneuvers aspiring pilots must master to obtain their license. She just as easily could be referring to her career and the circuitous route she took to achieve her dream of becoming a pilot. Ultimately, she would reach "the pinnacle," commandeering helicopters flying personnel to offshore oil and gas installations in far-flung locations such as the Philippines, Nicaragua, and numerous places off the coast of Africa, but it would take almost 20 years and there would be many detours along the way.

She discovered her passion for aviation at the age of 16 when her father's friend took her up in his "little fixed-wing and I thought, 'This is what I want to do. I want to fly.'" But she quickly came back to earth – literally and figuratively.

"At that time (mid-80s), neither Lufthansa[84] nor the German Air Force[85] would have taken women," says Nina. Taking into account the cost involved, the fact that her family had no background in aviation nor did it own a plane, Nina thought, "This is not going anywhere," and says, "I actually did forget about it for many years."

The memory of a car accident that left her hospitalized at age 15 and an older brother, who was studying to be a doctor with the military,

---

[84] In 1988, Evi Hetzmannseder and Nicole Lisy became Lufthansa's first female co-pilots. Twelve years later, they became its first female flight captains. Source: Lufthansa Group.

[85] In 2007, Urike Flender became the first female German Air Force jet-fighter pilot. Source: Holloman Air Force Base, New Mexico, USA.

coupled with her interest in science, convinced her that medicine was a viable career path. In reality, it would prove to be almost as difficult as aviation would be later.

Despite the fact that she was already working at a hospital as an assistant nurse during high school, getting into medical school in her native Germany was very difficult. Fortunately, an opportunity presented itself and she was accepted to attend Semmelweis University in Hungary. After a year, her realization that Germany had a surplus of doctors and a severe shortage of residencies – or what she refers to as "Grey's Anatomy positions" – led her to conclude her studies.

It took less time to realize business school "was not my thing," despite having received a scholarship. Looking for a profession that both interested her and would allow her to support herself, she realized her passions – science, medicine, and animals – were the prerequisites for veterinary medicine and spent the next five and a half years studying at the University of Berlin.

On her first break after years spent dedicated to her studies, she applied for a four-week internship at the brand new Two Oceans Aquarium in Cape Town, South Africa. As a certified diver, she thought, "I'll do this for a holiday!"

At the Aquarium, she met a South African veterinarian who offered her a position at his small animal practice in Cape Town, and she accepted. After a year's residency, she was told by the veterinary board that it would not recognize her German qualifications.

"Working for the year, I can tell you, veterinary science is not James Herriot," Nina says wryly. "Still, I would love to stay in Cape Town – friendly people, lovely weather, better opportunities than in Germany."

Not willing to let her years of study in Germany go to waste, she approached the University of Cape Town and was told she could do a post-grade in zoology with a major in marine biology.

"I got into the study course, got my degree, can pin it on the wall, and say that I've got my Bachelor of Science Honors master's in marine biology from the University of Cape Town in South Africa!"

While she was thrilled to have figured out a way to earn a South African degree, she still had to make a living and says the "easiest, best money" was working on the weekends as a tour guide "driving German tourists in the big buses and speaking my native language."

As predicted, "Once my degree was finished, of course, there was no job in marine," she says, but the travel company she worked for asked her to stay on and continue as a tour consultant and project manager for incentives and specialized tours. On occasion, it involved booking helicopter tours.

One day, her father, who "was a bit of a TV personality in Germany – so he had a few connections," called her with a request from DAV – *Deutscher Alpenverein* (the German Alpine Club) – the biggest mountain club worldwide. They asked him if Nina could put together some tours for them.

It was the catalyst for starting her own company as an inbound tour operator in southern Africa, something she would do for the next ten years. In 2004, in what would prove to be a serendipitous turn of events, Nina received a request from a group of Scottish golfers, who wanted to be flown by helicopter to the golf course and asked her to accompany them.

"I thought, 'I've never been in a helicopter.' I loved it!"

She sat in the helicopter talking to the pilots, who told her they had just come back from a whale count in Cape Point and how incredible it was. The same feeling she had at 16, going up in her father's friend's fixed-wing, came over her, "Here it is again – this is what I want to do," she says, the excitement still evident in her voice.

The pilots, one military and one civilian, explained that she could go straight into helicopter training at flight school without doing fixed-wing first. After fifty hours, she would have her private license. At that time, the cost was 120,000 rand,[86] "which was a lot of money," she says,

---

[86] At the time of this interview, the exchange rate was 1 ZAR = 0.0765 USD, 120,000 rand is ~$9,180.00 USD.

"but, I had worked for ten years. I had my little house, my business, my car, my toys, my surf ski, my motorbike. I was settled."

When the smooth-talking sales and marketing person of a flight school in Cape Town offered her a deal, she thought, "It just sounds too good to be true. I immediately signed up!"

She then met the flight instructor, Frank Peters-Hollenberg, a renowned Namibian-German pilot, who was among the first to conduct sea rescues. Conversing in German, they immediately hit it off, but she was taken aback when he asked brusquely, "Why do you want to do this?"

Nina told him she had always wanted to fly.

"Do it for fun, but keep your day job," Nina recalls him saying. It became clear from talking to the other pilots and trainers that there were no jobs, and she had the added strikes of being a female and a foreigner. "Just to let you know," Peters-Hollenberg said, then added, "but I'll fly with you any day."

And off they went in a little Robinson-22, two-seater, open doors and, once they got up high, he turned to her and said, "Okay, you've got control."

"And I have everything *out of control*," Nina recounts. "It was a complete nightmare. I thought, 'I'm going to kill both of us, right now, right there.' When we landed, I was nearly green in my face, close to being seasick. I thought, 'This was horrible. I never want to do this again.'"

But with five hours already bought and paid for, and the realization she may have just saved her own life, she marched into Peters-Hollenberg's office and said, "Okay, Frank, what times are available?"

And he responded, "Let's go ahead and do your private license."

Because she was still running her tour business, it took six months, at the end of which time, she told him she wanted to go commercial. "I was fascinated by a life of flying. *This was it.*"

Again, Peters-Hollenberg was blunt. "Nina, you're not going to make any money. The military guys are taking the good jobs. You're not South African, you're not in your 20s – 'by then, I was in my mid-30s,' she says – and there are not many females out there."

In fact, Nina was the only female student at the flight school at that time. There was another woman, who had just finished South African military service and who flew the rescue helicopter, whom Nina looked up to. Although she was very friendly when they met, she cautioned Nina, saying, "It's hard to stand your ground."

Nina remembers thinking at the time, "I'm pretty strong, I was a competitive swimmer and runner in school. At 1.76 meters,[87] I'm tall. I'm not one of these small little petite ladies. I can do that."

The first surprise came when she discovered there were no commercial helicopter courses offered to civilians and she would have to navigate the rigorous South African testing process on her own in order to earn her license. It took nearly two years of independent study and completion of the 200 flying hours necessary for a commercial pilot's license (CPL), something she describes as "an ongoing challenge," as flight schools opened and closed with regularity in South Africa – not to mention selling her house, her car, her motor bike and surf ski to pay for the hours!

"Luckily, I had my mentor, Frank Peters-Hollenberg," she says fondly. "Still, to the day, we stay in contact about my career. He loved my determination because he said everything was against me at times but I was so passionate, I just wanted to do it. He's been the person who has really helped me and guided me along, especially in the beginning." It is obviously a mutual admiration society as Peters-Hollenberg gave special mention to Nina as one of his "memorable students" in his 2010 memoir, *Helicopter Odyssey*.

"Frank's military, of course – everything in South Africa is connected to the military in the end," she says and he referred her to another ex-military pilot, Danie Terblanche, a trainer, examiner, and test pilot in Nelspruit, close to Kruger Park in the north of South Africa, with his own helicopter, who could help her complete her final hours of flying time.

"Maybe I never learned to fly the perfect traffic pattern," she muses, "but I definitely was flying underneath trees, and landing on dam walls, and landing on one skid – all of these goodies – because that's how it was there – a small airport in the bush. It was absolutely amazing."

She finished her CPL in 2006 under the guidance of Danie, whom she describes as "one of these wonderful people, who is impressed by passion and determination, and would go way beyond to support you, regardless of race, nationality, gender or age – as long as he was convinced that as one of his students you would fly safely enough to stay alive!" He again called some ex-military friends who were looking for a pilot that could start in a week's time and work during the fire-fighting season in KwaZulu Natal. There would be little to no money, but, as far

---

[87] About 5' 7 ¾"

as Nina was concerned, there was a huge payoff – a conversion on a Russian MI-8.[88]

"We only had two fires from June to October of 2006. I was sitting in the middle of nowhere. We flew 27 hours in total. We were on standby every day. I had my first little idea of how it was going to be and it wasn't how I imagined." But then she brightens at the memory, "It was great. I was suddenly on a real big machine, on a twin turbine. Turbine time is what you need as a pilot."

Once the fire-fighting season concluded in October, she went back to the western Cape and did her yearly Crew Resource Management (CRM) and her dangerous goods (DG) course. While there, she met an ex-fire fighting pilot who was running scenic charter flights over Victoria Falls in Zambia. He asked if she was interested in flying for him and she said, "Sure!" Two days later she was on a plane to Livingstone, although she confesses, "I didn't even know where it was."

In what by then was a familiar story, he told Nina he couldn't pay her much. Once again, the sacrifice was worth it to her. "I would get the hours and the ratings for three Eurocopter aircrafts – the AS350 ('the Squirrel'),[89] the EC130, and EC120."

Undaunted, she lived for a year without electricity or water in the bush in an A-frame hut without windowpanes, only mosquito gauze. In retrospect, she says, "It was the most beautiful time – I had elephants and hippos at my door, but spiders and snakes, as well."

The work permit allowed her to stay for two years, resulting in the ultimate payoff. "Fifteen hundred hours put you on the list – you exist as a pilot. Before, you're nothing, actually."

It was also where Nina had her "close call" as a pilot.

"I was flying Victoria Falls doing tourist and charter flights. In between charges, while hovering empty for refueling, I had a hydraulic failure, which ended with a complete write-off of the helicopter. Luckily, I did walk away."

Back in South Africa, Nina met the chief of aviation from the police department, who shook her hand and said, "Every pilot will have one accident; if you can talk about it, congratulations."

---

[88] A medium twin-turbine helicopter, originally designed by the Soviet Union, and now produced by Russia, the MI-8 is most commonly used as a transport helicopter, which can accommodate up to 26 passengers. Source: Russian Helicopters (part of State Corporation Rostec).

[89] Single-engine light helicopter capable of carrying up to four people. Source: Air Bus Helicopters, Inc.

In October 2008, she got the opportunity to start flying offshore. The company that she had done firefighting for had taken over a contract in South Africa and Namibia. For the next three and a half years, she flew the Sikorski-61 offshore to ships and platforms off the Eastern coast of South Africa, as well as Walvis Bay, Namibia, where she was based for six months.

Despite the grueling schedule – seven days on/one day off, then fourteen days on/two days off – the emergency phone around her neck 24/7 – and even with 1,500 hours, her lack of offshore experience resulted in a salary so low it made it necessary for her to work on the side translating hotel brochures into German just to make a living. Still, Nina calls it "a huge opportunity."

In September 2011, a recruiter for a multi-national aviation company went to Africa to find touring pilots for its Africa/Asia contracts, which involved working a rotation schedule. Her big break came three days later when she was offered the job, despite the fact that she didn't have an Airline Transport License (ATP) and was still flying on a CPL. Shortly thereafter, she found herself on a plane to Vancouver, Canada, for her first ever simulator training. Her colleagues – all men – were retired military pilots with 25-plus years' experience.

"Here I was, six years' experience, never been in a simulator, never flown an automated aircraft," she says, marveling at her good fortune. "I was thrown into the cold and deep water. It was beyond extreme, but I was determined."

She feels like her employer took a risk on her – not because she is a woman, but because she didn't have the background it was looking for – and has tried to show her appreciation by being an exemplary employee, using her off-time and personal finances to complete her FAA ATP in the US She says there has been an incredibly steep learning curve over the past four "really amazing, intense years," but that she has received unwavering support from her employer.

"There are lots of pilots who have more experience than I do. Most come from the military, they've been in the game 20-plus years, I'm not even 10 years in the game, but I'm playing the same field. I started pilot training in my mid-30s, so it's never too late."

In retrospect, while she says she wouldn't necessarily recommend that others take the circuitous route she took to become a pilot – instead, enlisting in the military or enrolling in a recognized training academy – she is disheartened by the lack of women coming up in the ranks.

"I think it's the effort you have to put in, the sacrifice in lifestyle is too hard. I have one female colleague who has a family and it's amazing how she can do it. I had to make the decision to say, 'Do I want to have a family or do I want to do this job?' because I wouldn't want to be away for six weeks, if I had children, and they can't come along. If I had been married and super happy, I might never have been a pilot. I'm very happy with the way things have turned out, but you have to make sacrifices."

Case in point: for the past year, Nina has lived off the coast of eastern Africa in a container camp surrounded by double barbed wire, often the only woman at a base with 85 men. While she acknowledges it is one of the ways in which her job is not always "women-friendly," that is not the part she finds most difficult to deal with.

In addition to flying oil and gas personnel to offshore installations, sometimes she and her colleagues are on 24/7 Medevac stand-by, and sometimes *only* on 24/7 Medevac stand-by, an exception which means little to no flying, "which for me is the hardest."

Not flying certainly doesn't mean she is not working. As flying base manager, she has numerous administrative duties managing the base, organizing work permits and authorizations for crew and aircraft, supervising travel for the pilots, as well as handling communications between the customer and client.

She is studying for her European pilot's license, and every year, pilots have their simulator test. If they were to fail a check ride, it would jeopardize their license and their career. "It's continuous studying, but maybe that's what keeps it interesting and keeps me challenged, while I got bored in other professions," Nina surmises. "Safety doesn't stop and technique doesn't stop. New instruments, new procedures, new things come up every day."

While she takes it upon herself to increase her knowledge and skills and earn additional licenses, she also believes, "You have to be incredibly pushy and driven. I've sometimes thought, 'I didn't do this right in the simulator and maybe I have forgotten it.' A man would never think, 'I haven't got it.' Women are very self-critical and the biggest problem in your career and your forward movement is that you actually pick out your own mistakes and you're the hardest with yourself."

That's not to say there haven't been some negative comments from men along the way, but Nina has found a way to deal with them.

"I've had guys say, 'We're not flying with a female.' And I look at them and say, 'Great, swim this way to the rig. It takes two days.' You have to keep it fun."

As captain, she also has the occasional situation where a male colleague is resistant to cooperating with a female. In Nina's mind, it is irrelevant; she would prefer they ignore the fact that she is female and just get the job done. She is equally dismissive on the subject of being a token female. "That may be thrown at you, as well as saying, 'Oh, you're nice looking and that's why you got the job.' You have to live with it."

She advises women considering joining the aviation industry to go in with their eyes wide open. "It's partly very lonely. It's non-stop having to stand your ground. A man doesn't trust a woman immediately like he does another man. You have to work for respect, continuously, every day, every new base; you don't have it given to you. You don't come with any kind of trust. You have to start all over again. But you can. There are mostly very good people out there, and, when you present yourself right, they will support you and you will become one of the team."

One of the few women Nina has encountered in her career is, not surprisingly, her recruiter and dispatcher/crew coordinator. As she points out, Human Resources (HR) is a female-dominated industry. "They have to work harder to get you in the door and present you as an option. There are females who actually do push for you, but you still have to bring the performance even more than the males very often."

And, indeed, she has brought the performance, as she believes offshore flying requires the top level of skills and expertise.

"We are continually in a hostile environment," she explains, pointing out that if a pilot is flying over land and something goes wrong, the pilot usually can find somewhere to land; over water, there are no options. Moreover, the helideck is a moving target and will often shift violently in the midst of a storm, especially on smaller ships. Pouring rain, crashing waves, and pitch-black nights can make the helipad barely visible, and create extremely tricky conditions for landing.

"It's much easier to land on a steady, well-lit, immobile runway. That's why we've got incredibly good training and why flying offshore is one of the pinnacles in aviation. No landing is ever boring but that's the excitement."

She then elaborates, "The fascination with helicopters is that they're not just used to fly from A to B. That's something that is different and challenging from flying a fixed wing. You have to work with them; they're a tool. Sometimes, we have to hoist things – materials, luggage, people – onto a ship. You have a lot of different variables. Not every ship is big enough to have a helipad, so we can't land or the helicopter is too heavy to land, so you have to hoist and sling. That's part of the job."

Being on 24-hour medevac stand-by presents its own challenges, as the pilots are required to go out any time day or night, in any weather, despite the fact that it's not something they do on a routine basis, which highlights the critical importance of keeping skills sharp through flight simulator training and regular night landing drills.

Having been in other industries that are not as safety conscious as the oil and gas industry, Nina marvels, "These are the highest safety standards in aviation that I know. I see a real professional, state-of-the-art environment, as far as it can get for civilians. We're playing in the top league and that's what I enjoy."

While she is appreciative of the industry's efforts to provide safe working conditions for its pilots, she is acutely aware of the "huge responsibility" she has for her passengers. "Currently, I am flying a Sikorsky S-76®,[90] and I have up to 12 people sitting in the back who trust their lives to me. You have to have a very disciplined lifestyle. I don't smoke, don't drink, I stay fit. You stay on top of the game as much as you can."

On the rare occasion when one of those passengers happens to be another woman, she says, "It's such a joy for me if I have a woman on my heli to fly around." She feels there is a sense of recognition that passes between them. "I think, 'Go, girl! I know it's not easy.'"

"If you want something, don't allow anyone to tell you it's not possible," she says resolutely. "I've been told so often, 'Oh, you won't be able to, and you can't, and you haven't got it.' It's never true. As I say in German, '*Geht nicht – gibt's nicht.*'"[91]

"I think I've gone all the way right to the top, actually. And I'm loving it still," Nina says with conviction. "Right now, I'm like a nomad who lives nowhere, but I love what I do," which brings to mind a quote from her friend and mentor. "Frank has a very good saying, 'If you don't live on the edge, you take up too much space.'"

---

[90] Twin-engine helicopter capable of carrying two pilots and twelve passengers.
[91] Slang expression, loosely translated, "Impossible isn't possible."

23	**Marni Zabarsky**
	First Female Saturation Diver, Gulf of Mexico

> **"Being a female, you've got to be different,
> if you want to have any longevity."**

❝ I loved the ocean, absolutely," says Marni Zabarsky of growing up in Massachusetts on the eastern seaboard of the United States. "We had moved from Worcester to Cape Cod. I loved the water. I was always in the ocean and then, at 15 or 16, I had this horrific dream.

"You know you have these dreams where you're halfway between being awake and asleep. This was a really vivid image of a shark – a big, ole, great white. I looked over and there it was and it just chomped me in half. I literally fell out of my bed I jumped so hard. I was scared to death of swimming in the ocean after that. I had nightmares about sharks."

Once that seed of fear had been planted, Marni became convinced, because the dream had taken place in the ocean, that she would die in the water. "Living there, that was it, that was how I was going to go and *naturally...*" she drags out the last word, "I ended up working under-water."

Wanderlust took hold early and, even before college, Marni says she was "scraping by just to travel." After graduating in 1994 from Assumption College, a small, private, Roman Catholic liberal arts college in Worcester, Massachusetts, with a bachelor's in communications, and a minor in political science, she says, "The bills started coming in and all I wanted to do was travel."

She began writing for the *Worcester Phoenix*, a local spin-off of the *Boston Phoenix* magazine, and other publications, writing travel pieces and taking photos from various locales – all to fund her passion for traveling.

Then in what would prove to be a fortuitous event, she says, "Puerto Rico happened." She met Irving Greenblatt, a publisher for Gannett newspapers at the time, who had turned an old plantation home into the Casa del Frances hotel on the island of Vieques on the northeastern coast of Puerto Rico.

"He used to hire 20-something year-old kids to go down and work there, be the wait staff. I got wrapped up in that. We weren't charged room and board, there was no overhead, and yeah," she admits, "I just wanted to go to Puerto Rico."

At the time, the United States Navy had a 23,000-acre facility on Vieques where it conducted military exercises.[92] One afternoon at a restaurant / bar on the beach, Marni began talking to some Navy divers, who asked if she liked diving. Although she had her scuba diving certification, she replied that she rarely went diving, preferring to snorkel and swim instead. The Navy divers told her she could make good money diving and suggested she get into commercial diving. "I didn't have a grasp on what it was, though," Marni says.

A few months later Hurricane Luis was headed for the island (although it veered in another direction), and she returned home to Worcester and took a desk job in the human resources (HR) department at the US Medical Center – soul-crushing to the free-spirited traveler. "Sometimes I would just stare out the window and think, 'What am I doing?'"

Spying a copy of *Scuba* magazine on a colleague's desk one day, she grabbed it, and saw an advertisement for Ocean Corporation's commercial diving school in Houston, Texas, and decided to enroll despite the fact that, "I, literally, had no clue what I was getting into. Not a clue."

Given the nightmares about sharks that haunted her dreams and the *Jaws* movie franchise, much of which was filmed in her home state of Massachusetts – "*Jaws* ruined everybody" – becoming a saturation diver wouldn't seem a likely career choice.

"Money is a great motivator," she says dryly.

---

[92] Source: Naval Facilities Engineering Command Atlantic (NAVFAC Atlantic)

When she first started diving commercially, one of the supervisors at the company she was diving for told her she was too pretty to be there and suggested she go back to Boston and marry a rich doctor. Marni told him he and her mother could sit down and talk, but until then, "I'm still here." He asked her how much money she thought she was going to make, and she responded, "A hundred grand." He told her it would never happen.

"Every year I just wanted to send him my W-2!" Marni says gleefully.

After enrolling, she spent the next seven months in dive school and, although she says, "Ocean Corp. is a great school; I love it," (she is currently on the advisory board), in her opinion, that was too long. "You can only do so much training in the classroom, and practicing in water tanks. It's not the real deal. You learn more just by slapping on a red hard hat, which is the international sign for you're new, look out, you don't know what's going on."

She explains that experience is gained "piecemeal as the supervisor kicks down dives to "tenders" – apprentice divers – and you work your way up the ranks. The most important thing you need to know is how to run a chamber because that's what you're going to be doing a lot. How to run decompression on divers. How to rig things on deck. How to be a resourceful tender. Knowing how to set things up – from the smallest thing like putting a socket on an impact wrench to rigging up tools and equipment – without sending them down to a diver to fail. The communication chain has to be solid."

Marni began diving for Global Industries in 1996 at the age of 24 and would spend the majority of her offshore career with them.

"Diving was the first job – clearly, a career – that held my interest. Although I never went into it career-minded; it was just a job I wanted to get paid lots of money to do," she says laughing. "It surpassed my three-month mark of losing interest and moving onto something else. Fourteen years later..." she trails off, still sounding somewhat mystified that her career spanned that length of time.

In the end, her career had staying power but she said she essentially had to fight the powers that be to get into saturation diving, where women divers are a rarity.

"They're very nervous about putting a woman in sat. Executive Vice President Jim Doré pulled me into his office. I asked him, 'Do I have the opportunity to earn my right to go into sat here or do I have to look elsewhere?'

"He said, 'You look like my daughter. I don't want you to go into sat and I don't want to send you overseas, but I can't stop you.'

"He held out a couple of evaluations from senior supervisors that I was diving for and with – I had no idea they wrote these evals – and there were high recommendations to put me in and keep me up with the peer group. And so he said, 'I can't stop ya.' I eventually got in and was very, very humble.

"I shrunk back down into a brand new red hat and felt the weight, the gravity, of what I had just done. I got into sat and now all eyes and ears were on me from the chamber to the water and it was nerve-wracking.

"That first sea dive – phew – I was almost scared to move the boat around. But I held my own. Hand-jetted in sugar sands[93] for four hours or five hours. I mean, I had to get it done. My bell partner Kevin Erickson who, from that point forward, was my bell partner for many years, said, "You knocked that out."

Nonchalantly, Marni responded, "Oh, I don't know, I did what I was supposed to do, and I'm really hot and thirsty." She admits she was exhausted but says, "I didn't want to fail."

She has retained that same humbleness she had when she first got into sat. "There was a diver who worked for Global for years before I did named Irene, who really *should* have been the first female in sat. Big German chick; she could handle her business," [make note: this is high praise from Marni]. "She definitely paved the way for the females that came in behind her. Irene was so senior to me; she intimidated the hell out of me. I respected her for having been there as long as she was and for being respected by the other divers."

Saturation divers live in a pressurized chamber for a period of up to 28 days – not something for the claustrophobic – but which never bothered Marni. (In fact, she says she had never had a panic attack until she was the passenger in a car on the Houston freeway!)

The chamber is pressurized to the same pressure as the depth of the sea where the sat divers will be working. The chamber stays at the surface and is attached to the offshore installation. When it is time to begin working, divers are transported underwater by a diving bell, which has been pressurized to the same depth. Remaining at pressure for the duration of the 28 days means the divers can live and work without

---

[93] Hand-jetting refers to using a metal pipe in the shape of a "T" connected to a high-pressure hose to blast through the fine ("sugar") sand on the ocean floor to dig a trench, typically to bury a pipeline.

having to decompress until their hitch is over, with the goal of avoiding a dangerous and potentially fatal condition known "the bends,"[94] in which bubbles of inert gas form in the bloodstream and body during rapid ascension and decompression.

While they are working underwater, divers are breathing "mixed gas" (helium and oxygen), which saturates their bodies and enables them to dive far deeper depths than they could on oxygen alone. It also causes Mickey (Minnie?) Mouse voice! "The first thing you do when you're in the decompression chamber is press your face to the portal, try to get a signal on the cell phone, and call someone – and they can't understand a word you're saying!" Marni says gleefully.

Sat dives typically vary in depth from shallow (as little as 100 feet) to 1,000 feet seawater depending on the nature of the work being done. Marni's deepest dive was 450 feet working in the West Delta block in the Gulf of Mexico, cutting out "coflex" (hose),[95] setting sandbags – "It was beautiful with unlimited visibility. Those dives are rare."

At the end of the 28 days underwater, divers only have to decompress once as they are brought back to surface pressure; however, for every 100 feet underwater, they require one day of decompression time, plus one extra day. (For example, a dive at 300 feet would require four days' decompression time.) "Essentially you get through decompression watching marathon sessions of TV series and movies you would never watch at home... and ordering cups of ice."

Craig Milburn and Tony Theriot are the two supervisors who gave Marni the high evaluations and with whom she credits for getting her into sat. Both men would become mentors and, while she praises them equally, they took different approaches with their mentee.

"Craig Milburn; call him Smoker," she says. "That's his nickname. He doesn't smoke, though. Smoker was more hands-on. I would never not credit that man for taking me under his wing. I've yet to meet his match as far as knowing how to run the job from diving to rigging to heavy lifting to getting stuff done.

"Tony was more of a 'get the hell out there and let's see whatcha got.' Tony benefitted from all the training Smoker gave me and made me prove myself.

"Those guys kept me glued to their hips when it came to running jobs and I learned a hell of a lot."

---

[94] Also known as nitrogen narcosis and decompression sickness.

[95] Industry slang for Coflexip® hose, an example of a name which has become a generic for reliable high quality flexible steel pipes. Source: Technip

She also gives high praise to her bell partner of many years, Kevin Erickson, who was with her on that first dive and whom she says to this day is still her buddy. "He was diving into his 50s with me. He's an old school hardcore. Excellent diver. He was really good with the tips and tricks some divers [keep to themselves] and don't pass down. And he would certainly hand me my ass if I was getting lazy about something!

"None of the guys I've mentioned held back telling me when I was screwing up and I used to say, 'You didn't yell at her [another female diver] in the water and look what she's doing.'"

Their response: "You're different."

Whether or not that was always a compliment is debatable Marni says, but, "Being a female, you've *got* to be different, if you want to have any longevity out there."

On the Ocean Corp. website under "Achievements," it says, "Ocean Corp. Alumn[us] Marni Zabarsky becomes first ever female Saturation Diver – 2001."

"In the Gulf of Mexico,"[96] she clarifies.

When she made her first sat dive, not everyone saw it as such a momentous occasion. Marni recalls the lead tender, who was in charge of all the "tenders," telling her she "sucked" and she "sucked twice as bad" because she was a "red hat" *and* a woman.

She says she just let his comments roll right off her back, like water off a duck, and didn't give them a second thought. "And that's pretty much the attitude you have to have."

When the job ended after a few months, she says, "That guy shook my hand and told me he would recommend me for any job he was on. It's just a matter of proving yourself and not having a chip on your shoulder," she concludes.

Marni's attitude served her well, not just with the lead tender, but with other divers she would encounter in the real world – and the virtual one. After becoming a sat diver, she joined an online dive forum and on one hand, received accolades from male divers congratulating her on her accomplishments and, on the other hand, "all these other guys were saying, 'She's not the first female... she can't do her work... they're just letting her in because she's a woman'... and, you know, just talking a bunch of crap about ya. I'm known in the industry because I'm a female

---

[96] In 1979, Susan (Sue) J. Trukken became the first female saturation diver while at the United States Naval Experimental Diving Unit (NEDU). Source: Navy Diver Foundation

that made it into sat and I stayed in sat. It wasn't a one-week deal; it was years" – fourteen, to be precise.

Marni says she always downplayed "everything."

"It wasn't something that I felt deserved attention because I was in there with the most senior guys – excuse my mouth – they worked their asses off for years, they missed their children growing up. There are serious rock stars in there. I was with some of the best divers I've ever met and it is just not something that I wanted to be the poster child for. It was always hard for me to accept or take a compliment or an "attaboy" for doing anything."

Being surrounded by that level of skill and talent, Marni says she always considered herself a "mediocre" diver (no doubt she is being overly modest), but that she did well at her job – and stuck with it.

She says the recognition was uncomfortable for her until she retired at the age of 38 after 14 years of diving. "I got out of the tank in 2010 and got into sales and realized how many people knew my name."

She also discovered there were still people that, for whatever reason, were unwilling to give her credit where it was due. When she went into project management, she encountered a salesman who pulled her aside and told her his colleague said although she had gotten into sat, her bell partner carried her the whole time. Marni burst out laughing.

"One bell run is a lot to carry somebody so I couldn't imagine being carried by the same bell partner for years. I thought to myself, I'll have to call Kevin and tell him, 'Hey, Kevin, so-and-so says you carried me my whole career.' He'd be like, 'You're too heavy to carry!'" she says, continuing to laugh at the absurdity.

While she never got to dive with any other women sat divers during her career, Marni did get to work for a woman after she left Global in 2008 and went to work for Julie Rodríguez , then-president & CEO of Epic Divers & Marine, which had become a subsidiary of Tetra Technologies, Inc. in 2006.

"It was pretty cool for me to go work for a dive company that was run and operated by a woman. She's no joke; *she is handling her business*. She's very successful in her industry." (So successful, in fact, that when Epic sold its assets to Tetra in 2006 it was a $50 million dollar cash transaction.[97])

Just as her bell partner, Kevin, generously shared the tricks of the trade with her all those years ago, Marni, in her position on the Ocean

---

[97] Source: Legacy Capital

Corp. advisory board, now offers her hard-fought wisdom and insight to young female divers – something she is uniquely qualified to do.

"They have their share of females," she says referring to the four or so women in each graduating class. "I'm always happy to speak with them. I let them know what they're getting into. I tell the girls, you're going into a male-dominated industry; there's no question about that. You're not going to go out there and change the mentality; you're not going to go out there and change the structure [of the work environment]."

She advises the female divers to ignore guys who may be less than welcoming – "especially if you work circles around them – phew," but she also cautions against others who have low expectations of women.

"You can't pay any attention when you make a dive and you get down bottom and all you do is tie off a piece of rope and when you come back up, the company rep says, 'Hey great dive; you did a great job.' You feel good about yourself but you didn't do anything different than the guy before you or the guy after you. I always tell a girl, they don't expect that much of you; it should be an insult. You should be like, 'Of course, I did a good job; it was tying off a rope. What do you expect?'"

Having had her accomplishments "casually discredited" at times, Marni says the biggest compliment is when a man tells her he wants his daughter to meet her and talk to her about her diving career. She hopes to be able to "watch those girls grow up and see one of them go offshore and conquer it."

Going from diving to project management was a natural progression Marni says, but she found herself managing her fellow sat divers and fielding complaints about things like getting plastic forks offshore instead of silverware. "Glorified babysitter," she grouses. "I absolutely pulled out my hair!" That wasn't the worst of it, though.

"Going 'into the beach,'" as it is referred to in the industry, "was a rough transition. You go from diving, where you're on rotation, 30 days on, 30 days off, to an office job. I destroyed about three alarm clocks in the first few months. Office life was not something I was ever hip to. I hated it." Shortly thereafter, she became pregnant with her daughter – "a happy welcome to my life" – never to return offshore.

She hasn't found the satisfaction working onshore that she did offshore and many times has thought, "I'm just so done with this industry. I'm going to get out and do something completely different, and then it just pulls me back in. It makes no sense."

After spending over two years in project management, she decided to transition to business development, which Marni says she found to be "far more accommodating to a new mother."

She has been with Magellan Marine International since 2012 and says, "Now I'm selling to people like me who are retired from offshore diving that are technical experts that work for Shell and Chevron and Exxon and sit in-house and are technical advisors when diving projects come up. It's all related. There's so much opportunity in this industry." And maybe that is what, in the end, keeps her coming back.

"Right now it's important for me to instill in my daughter that she shouldn't see herself as being limited because she's a girl. I told her, be thankful you were born in the United States of America. This country has allowed me the opportunity to work shoulder to shoulder offshore – sweating, bleeding, and tired – next to a guy and get paid just as much."

In the past, Marni has been reticent to talk about her career but now says, "A lot of the pride I feel talking about any of this is that my daughter will be able to see that there are no excuses. You can do what you want to do."

<table>
<tr><td>

24

</td><td>

Alyssa Michalke
First Female Commander
Texas A & M Corps of Cadets

</td></tr>
</table>

"I refuse to let anybody outwork me."

Alyssa Michalke has come full circle and, in doing so, she has made history. The 21-year old was born in College Station, Texas, and, while she only lived there the first year of her life, she returned in 2012, having been accepted as a freshman at the prestigious Texas A & M University. At the end of her junior year in 2015, while pursuing a major in ocean engineering, as well as a certificate in project management, Alyssa was named the first female commander of the Corps of Cadets in the school's 139-year history.

Although Alyssa's mom, Nicole Michalke (née Bludau), graduated from A & M with a bachelor's degree in computer science and a master's in agriculture economics, Alyssa didn't always aspire to attend her mom's alma mater. In fact, at one point, she didn't have aspirations to attend college at all.

Alyssa's dad, Rodney, who earned his welding certificate from Texas State Technical Institute (TSTI)[98] in Waco, Texas, used to come home from working his hitch as a welder on platforms in the Gulf of Mexico (GoM) during the '90s and capture the imagination of his young daughter with tales of his offshore adventures.

"He told me stories about fishing and catching sharks and all the food they would eat, and going out on a helicopter for two weeks at a time, and I was thinking, that would be cool, I could go to a trade school for two years and become an underwater welder."

Her dad told her it was her life and she should "live like you want." Her mom, who was valedictorian of her high school class, and later

---

[98] In 1991, the school was renamed Texas State Technical College (TSTC).

maintained a 3.999 GPA at A & M before going to graduate school, had other ideas.

"She said, 'Alyssa, you're going to be valedictorian of your high school and you're going to go to trade school for two years and become a welder? No, I don't think so. You need to apply yourself a little bit more.'"

Her mom, whom Alyssa credits with being "extremely smart," told her she probably ought to go to 'college college,'" and Alyssa says, "The mature side of me said, 'She's probably right.'"

Despite their differing perspectives on that particular matter, Alyssa calls her parents her "rocks," and says they were her role models and mentors long before she went to college.

"They are really hard-working and successful, and supported me and my younger brother, Jacob, who is now a high school senior, throughout our childhood and teenage and high school careers. They set me up for my first successes," which would include excelling both academically and athletically in Schulenburg, the small town in central Texas, where her parents moved about a year after Alyssa was born.

"I was really good at math and science courses early on. I like challenging myself. It's one thing I do best."

As much as she gravitated toward those subjects, her academic choices were limited. After taking algebra I as an eighth-grader, she looked forward to taking calculus in high school but discovered that high school math classes at her school didn't extend beyond pre-calculus.

"My senior year, I didn't even have a math class."

However, the seed had already been planted. "I really enjoyed the challenges of my pre-calculus class taught by Mrs. Lori Kallus and my physics class with Mrs. Mercy Silcox during my junior and senior years of high school. I liked the problem-solving aspects of them and I was actually pretty good at them."

Alyssa credits those two teachers, in particular, for pushing and challenging her, and believes they played a part in turning her focus toward STEM subjects and the engineering field by confirming her aptitude at solving complex problems.

"That career path would always be interesting to me. I wouldn't be sitting in a cubicle for days on end, crunching numbers." As an avid outdoorsman, "I couldn't be cooped up in a big city office all day, doing the same thing. There would always be a new challenge, something to keep me on my toes."

A multi-sport athlete who ran cross-country, played basketball, softball, and golf in high school, Alyssa had an alternate career path in mind.

"I love coaching and mentoring and teaching. I thought it might be cool to be a high school math or science teacher and a basketball, softball or golf coach – or a college coach, if I got to that level. It's two sides of the spectrum that are not related." She reiterates, "I knew that I wanted a challenging career and to commit myself to life-long learning and bettering of myself."

Ultimately, between the guidance from her teachers and the inspiration from her dad's career as an offshore welder, Alyssa decided to major in petroleum engineering. In a rare disappointment for a young woman used to excelling, she discovered that being valedictorian of her high school class, as well as an award-winning athlete with automatic admission to A & M, didn't guarantee her choice of majors.

"Funny story, actually," she says in her clipped, military-inflected speech. "I really wanted to major in petroleum engineering but, because I was behind the power curve, I got into ocean engineering. I have loved all of my classes so far. I'm really looking forward to the oil and gas industry. Even if I had gone into petroleum engineering, I still would have wanted to go into offshore oil and gas and not stay land-based the entire time."

Because of a curriculum change, she also was not able to dual major in ocean and civil engineering, but instead chose to pursue a certificate in project management to complement her major in ocean engineering.

At the same time, she decided to go into the Corps because she wanted to challenge herself just as much as she had in high school – academically, athletically, and socially – and to keep an equally busy schedule in pursuit of her goals.

Despite her achievements putting her in the public eye, Alyssa says there is something people don't know about her. "I'm actually an introvert. I love my alone time – reading, hunting, running, fishing, working by myself in my room – so getting outside of my comfort zone has been a huge challenge for me throughout my life."

She tells the story of her high school cross-country coach, Richard Hoogendorn, who constantly told the athletes as they were running up and down hills or had just finished working out in the weight room "until our legs were dead," [affects a male accent], "'You know, if you wanna get good, you gotta be comfortable with being uncomfortable, you gotta be breathing hard, sweatin', actin' like you're gonna throw up, in order to get good.' That was always great to hear at 5:30 in the morning," she says laughing. While not exactly what she wanted to hear at the time, his words, "You gotta be comfortable with being uncomfortable" stayed with her and continue to motivate her to this day, as she pushes herself to move beyond her comfort zone.

"When I mentor other students, I tell them, 'Look, I'm a small-town kid from Schulenburg, Texas, who never even dreamed I would be in this position. I can definitely tell you, if I had not pushed myself, gotten outside of my comfort zone, I wouldn't be here, because I'm an introvert.'"

She credits the Corps and its well-defined, four-year leadership development program with giving her the insight to identify her strengths and weaknesses, build on those strengths and "attack" the weaknesses. The result has been an immense amount of self-confidence.

"I am very confident in myself and my abilities. I have an exceptionally strong work ethic; *I refuse to let anybody outwork me*. My dad told me from a very early age, 'Alyssa, no matter what position you want, there are probably 100 other people who want it as well. They want to be the starting point guard, they want to be the Corps Commander, they want to be that top executive in an engineering firm. If you're not outworking them on a day-to-day basis, they're going to get it instead of you, so you have to have that work ethic and push through.'"

The Texas A & M Corps of Cadets recognized those characteristics when it chose her to become the first female student to lead the Corps as Commander.

She acknowledges that it is a "huge honor and privilege," but is quick to give credit to the women who came before her who made great strides, not just in the Corps of Cadets, but at Texas A & M university as a whole, and there is no doubt she includes her mother in that group. Alyssa says it is because of their hard work, determination, and perseverance that she has the privilege of serving alongside some of the best leaders that the Corps and Texas A & M have to offer. "It's a great opportunity for me to develop myself and those around me." She can't resist adding, "I'm looking forward to all the *challenges* that this year is going to offer."

Because she is the first female commander, one could assume those challenges would include finding mentors, but she says she has had "exceptional mentors" – a lot of them male, something her friends find "weird" – throughout her time in the Corps. She calls Sean O'Connor, a chemical engineering major who was a 2014-15 outfit commander, one of her best mentors.

"He was one of my direct superiors when I was a freshman and sophomore, held me to extremely high standards, challenged me on a daily basis, and helped me develop as a leader." Now that he has completed his four years in the Corps, Sean is finishing his degree as a "regular student," according to Alyssa, walking around campus "with long hair, in shorts and t-shirts." As someone who has experience in the Corps, but is now a neutral third party, she finds his input valuable.

Not all of her mentors have been male; she cites Morgan Cochran, Bachelor of Arts, sociology ('15), now a Second Lieutenant in the US Army, as "one of my great female mentors throughout my sophomore and junior years, another great role model and leader to look up to. She provided insight on some decisions I had to make. As a sophomore, you're only responsible for two to six people, but in my junior year, I was responsible for about 2,400. That's a huge jump and she helped me develop that larger scope."

Echoing what so many successful women in the industry have said, Alyssa reiterates that mentoring often begins at home and gives her parents full credit.

"Again, I would be remiss, if I didn't talk about my parents. They work extremely hard, have exceptional characteristics and traits, and have supported me throughout every single endeavor I ever wished to pursue, no matter what it was. I can definitely say without a doubt, the characteristics and traits that I have developed and that have made me successful, I learned from my parents."

A popular topic of debate in the media is whether women can "have it all," and there is no way Alyssa could have it all – academic success, athletic achievement, Corps leadership, and a personal life – without learning the fine art of time management, a realization she came to fairly early on.

"In high school, I learned, if I wanted to continue to be good at basketball and softball and golf, and be in the National Honor Society (NHS) and be valedictorian, and have a part-time job, and hunt and fish on the weekends, I was going to have to learn to manage my time; otherwise, I was going to have to give up something.

"Now, I live and die by my planner. I have an Apple iPhone that I put everything into. I even take a little 30-minute nap every day, and I build that into my schedule."

Alyssa shares an analogy given to the Cadets by their commandant, Brigadier General Joe E. Ramirez, Jr. ('79), that resonated with her. "He said every one of us is juggling these balls – glass balls and rubber balls. The glass balls are tasks that have to be done by today – immediately – because, if you don't and you drop them, they're going to break and your chance is gone. Then, you also have these rubber balls that [represent tasks that] need to get done eventually, but not necessarily right now, so you can drop one of those and it will bounce, and you can pick it up the next day or the next week."

She says the analogy helped her prioritize and plan, which are things she has learned to do well, not just for the week ahead, but for several weeks out. "Nothing sneaks up on me," she says, "and I'm not caught unprepared."

The Corps' overarching vision, she explains, is to develop well-educated leaders of character who embody its five core values of honor, integrity, discipline, selfless service, and courage, in addition to being highly sought after, academically successful, and prepared for the global challenges of the future.

"We have a different mindset this year of very purposeful and professional training because we're not just producing leaders for our nation's military, but also for the private sector. Only about 40 to 45 percent of our cadets actually go into the military. Not all of us are going to Afghanistan or Iraq to fight terrorism anymore. The majority of us are going to lead the nation in the corporate world, so we have to make sure our training reflects that."

While there is no military obligation, the Armed Forces know a leader when they see one and, not surprisingly, have their sights set on one of the Corps' best and brightest.

"I have a US Marine Corps colonel trying to convince me to go into the Marine Corps. I said, 'Sir, if I serve for four years, I'm going to have to go back to college to relearn all the stuff they're going to be doing in the oil and gas industry because they're advancing the technology and the systems at such a rapid pace.'"

She does admit entertaining the idea when she was younger.

"When I was growing up, I always played Army. Every time Halloween came around, I was a soldier or hunter, so my parents thought I might enter the military. I got to A & M and I realized that's not where God is calling me to go. He's calling me to go into the private sector. That's what I think His plan is for me. Until I think He's calling me otherwise, I'm going to stay on the private sector track," she says resolutely.

Ultimately, whatever she ends up doing, Alyssa knows it will have to challenge her mentally and physically. "Commuting to work by helicopter would be cool. I'm kind of an adrenaline junkie."

♀ ♀ ♀

Since graduating from Texas A & M, Alyssa has been working in the manufacturing industry, focusing on process improvement and operational excellence. She and her fiancé, Mitchel, will be married in summer 2019.

Afterword

A *Forbes* article, published February 1, 2019, stated, "At a recent event hosted by the International Association of Drilling Contractors (IADC) in Galveston, Texas, an energy company manager told the room that 'offshore rigs and makeup just don't go together'" (Spitzmueller and Brown). Apparently, he is unaware that they *have* gone together from the beginning. It would have been an opportune moment for the women in the room to stand up and say, "You're right – I didn't wear makeup when I started working offshore in the '70s… " or "Actually, I did wear makeup when I worked offshore in 1973… or 1987… or last week." But were there any women in the room to refute his comment?

If not, shouldn't this be when men, those who are allies and advocates for women, those who disagree with this kind of misogynist thinking, stand up and say something? Would it be awkward? Would it be uncomfortable? Not any more so than it is for the women on the receiving end of those types of comments. It is in these moments that we move from Time's Up[99] to "step up." It is how we solve this issue *together*.

---

[99] Time's Up Now is a 501(c)(4) social welfare organization working to create solutions that cross culture, companies, and laws to increase women's safety, equity, and power at work. Source: Time's Up

Acknowledgements

In memory of William Russell Mitchell who told me all those years ago, "You can be whatever you want to be," when I first dared to say I wanted to be a writer, long before I ever dreamed of being an author.

My heartfelt thanks to...

Jill West, my longtime friend and landman extraordinaire. This book would never have been written if she hadn't given me my first land job, which was my entrée into the world of oil and gas.

My son, Calvin Ponton, III, who helped with the early research of what would eventually become this book, and who is a gifted writer in his own right as well as being a photographer and videographer. Thank you for showing me what true courage is.

Cal Ponton, my former husband, close friend, and veteran of the industry, for his unwavering support and willingness to always answer my industry-related questions.

My daughter, Kristin Pearce, whose confidence in me motivated me to keep moving forward until this book became a reality, and whom I refer to as "Superwoman" for a reason.

My "cousin," Jean Paris, for being such a gracious hostess the many times she welcomed me into her home when I was in Houston and for always having a supply of dark chocolate on hand.

Janet A. Doleh, my first reader, for her hard work and efforts to improve my writing and particularly for reining in my overzealous use of commas!

Martha Miller, my friend and fellow author (*Times New Roman: How We Quit Our Jobs, Gave Away Our Stuff & Moved to Italy*), who bravely led the way into the publishing labyrinth and generously shared her experiences and knowledge with me.

My best friend of too many years to count, Nicole Burkholder, for always being there "come hell or high water," as my mom used to say.

My close friends, Cynthia and Frank Esquivel, without whom I would have no social life!

My family, especially Rod Van der Voort, and friends for never getting tired of hearing me say, "I'm working on the book," or at least not expressing their annoyance out loud.

Mi amiga Angela Levy for the love and laughter she brings to all our lives. How can one woman possess so many talents? #sheisanengineer

Dr. Mary Ann Tétreault, who passed away during the writing of this book, but who was supportive from the beginning and generously introduced me to the G2K group at Columbia and via e-mail to Sara Akbar, former CEO and co-founder of Kuwait Energy, who shared her incredible story for this book.

The families of Sarah Darnley, Dr. Zara Khatib, and Margaret McMillan for being willing to share their memories of their loved ones who are no longer with them. May each of these pioneering women rest in peace.

My "dream team" – Katie Mehnert, CEO of Pink Petro, for giving me a platform; Paula Waggoner, the Energy CFO, for her incredible generosity and impressive contacts; Kimberly Smith for her enthusiasm and encouragement; and Peggielene ("Nana") Bartels, Jo Ann D. Paul, and Carol Preslar for "praying without ceasing."

My friend, Karen Cawood, with whom I share many fond memories from our days in Dubai (the high point of which was the births of our sons) and also for her expert transcription skills, which saved me from having to transcribe - the bane of my existence as a writer!

My friend and mentor, L.B., for dozens of moments of inspiration.

At long last, here it is! I hope in some small way I have helped create awareness of a missing part of our history and honored the women who have not received the credit they so richly deserve for being instrumental in creating that history.

If there is anyone I have forgotten, please forgive my faulty memory and know that you are in my heart.

And, not least, to all my "sisters in oil" who are breaking the gas ceiling every day!

♀ ♀ ♀

Bibliography

Foreword

Alagos, P. (2015, March 11). Experts say energy sector hiring not hampered by oil price drop. *Gulf Times*. Retrieved from https://www.gulf-times.com/story/430490/Experts-say-energy-sector-hiring-not-hampered-by-o

LeanIn.org and McKinsey & Company (2015). *Women in the Workplace*, pp. 1-30, p. 3. Retrieved from https://womenintheworkplace.com/2015

Preface

Stonington, J. (2011, June 27). Boys-only boards: where women aren't at the top. *Bloomberg Businessweek*. Retrieved from http://www.nbcnews.com/id/43526947/ns/business-us_business/t/boys-only-boards-where-women-arent-top

IHS Global, Inc. (2016). *Minority and Female Employment in the Oil and Natural Gas and Petrochemical Industries 2015-2035*, pp. 1-162, p. 1. Retrieved from https://www.api.org/~/media/Files/Policy/Jobs/16-March-Women-Minorities-Jobs/Minority-and-Female-Employment-2015-2035.pdf

Accenture (2018). *Gender Diversity Study*, pp. 1-23, p. 9. Retrieved from the Petroleum Equipment & Services Association website https://pesa.org/wp-content/uploads/2018/04/PESA-Gender-Diversity-Report-April-2018.pdf

Oil & Gas UK (2017). *Workforce Report 2017*, pp. 1-10, p.10, figure 10. Retrieved from https://oilandgasuk.co.uk/wp-content/uploads/2017/10/Workforce-Report-2017-Oil-Gas-UK.pdf

PwC (2015). *The female millennial: a new era of talent*, pp. 1-31, p. 24. Retrieved from https://www.pwc.com/jg/en/publications/the-female-millennial_a-new-era-of-talent.pdf

Chapter 1 – WOW – Women on Water

Zonn, I. S., Kosarev, A. N., Glantz, M., Kostianoy, A. G. (2010). *The Caspian Sea Encyclopedia*. Heidelberg, Dordrecht, London, New York: Springer. (*Bibi-Heybat Bay is now Baku Bay | Ilyich Bay field.)

Aliyev, Z. (2011, Sept-Oct). The Art World of Maral Rahmanzadeh. Visions of Azerbaijan, pp. 76-84. Retrieved from http://www.visions.az/en/news/332/9ed2eaee/

Nazarli, A. (2015, November). Maral Rahmanzadeh: A Pioneer of Art in Azerbaijan. Azer News. Retrieved from https://www.azernews.az/culture/88023.html

Austin, Diane E. (2006). Women's Work and Lives in Offshore Oil. *Research in Economic Anthropology, 24*, pp.163-204 (p. 173). Retrieved from https://arizona.pure.elsevier.com/en/publications/womens-work-and-lives-in-offshore-oil

No byline (1973, September 27). Women on offshore rigs. *Lebanon (Pennsylvania) Daily News,* p. 39. Retrieved from https://www.newspapers.com/image/5442613/?terms=first%2Bof%2Btheir%2Bsex%2Bto%2Bwork%2Boffshore%2Bin%2Bthe%2BGulf%2Bof%2BMexico

Sterba, J.P. (1975, December 30). Amid the Grime, A Woman Earns A Job Offshore. *New York Times*, p. 20. Retrieved from https://www.nytimes.com/1975/12/30/archives/amid-the-grime-a-woman-earns-a-job-offshore.html

No byline (1977, July 14). *Stevens Point Journal*, p. 6. Retrieved from https://www.newspapers.com/image/251466173/?terms=Kristin%2BLovelace

Japenga, A. (1979, June 17). It's lonely topside for Heidi's sitters. *Los Angeles Times*, front page of The View section, cont. p. 14. Retrieved from https://www.newspapers.com/image/387452003/?terms=Kristin%2BLovelace; continued on https://www.newspapers.com/image/387452173/?terms=It%27s%2Blonely%2Btopside%2Bfor%2BHeidi%27s%2Bsitters

No byline (1978, July 12). Women joining the ranks of offshore drilling. *Paris (Texas) News*, p. 6B. Retrieved from https://www.newspapers.com/image/13797708/?terms=Women%2Bjoining%2Bthe%2Branks%2Bof%2Boffshore%2Bdrilling

Moore, R. (1984, November 8). What chance has a woman to work on an oil rig? *The Guardian*, p. 13. Retrieved from https://www.newspapers.com/image/259722263/

Canetti, B. (1982, October 18). Women proving themselves on offshore oil rigs. UPI archives. Retrieved from https://www.upi.com/Archives/1982/10/18/Women-proving-themselves-on-offshore-oil-rigs/3156403761600/

No byline (1978, July 16). Oil rigs and young lace. *El Paso Times*, p. 6C. Retrieved from https://www.newspapers.com/image/435401504/?terms=women%2Boffshore

Groden, C. (2015). Most powerful women: Europe, Middle East, Africa. *Fortune*. Retrieved from http://fortune.com/most-powerful-women-europe-middle-east-africa/ann-cairns-25/

No byline (2013, May 31). Mastercard's Ann Cairns: From 'one of the lads' on a North Sea rig to a City leading lady. *Evening Standard*. Retrieved from https://www.standard.co.uk/business/markets/mastercards-ann-cairns-from-one-of-the-lads-on-a-north-sea-rig-to-a-city-leading-lady-8639207.html

Kamenetz, A. (2010, July 1). Former BP exec Cynthia Warner left big oil for big algae – and she's not alone. *Fast Company*. Retrieved from https://www.fastcompany.com/1659080/former-bp-exec-cynthia-warner-left-big-oil-big-algae-and-shes-not-alone

Bridge, G., and Le Billon, P. (2012). *Oil*. Cambridge; Malden, MA: Polity Press.

Abdel-Magied, Y. (2016). *Yassmin's Story*. Australia: Penguin Random House.

Abdel-Magied, Y. (2017, September 28). I tried to fight racism by being a 'model minority' – and then it backfired. *Teen Vogue*. Retrieved from https://www.teenvogue.com/story/fight-racism-model-minority-yassmin-abdel-magied

Career Girls. [careergirls.org]. (2012, February 20.) Engineer/Executive: Working Offshore On An Oil Rig - Paula Harris Career Girls Role Model. Retrieved from https://www.youtube.com/watch?v=ms3tGXVOllc

No byline (1921, October 26). Diving Girl Spends Hours in Oil Well, Santa Ana Register, p. 9. Retrieved from https://www.newspapers.com/image/71410753/

No byline (1921, October 3). Venice Woman Diver Makes New Record, *Venice Vanguard Evening Standard, Volume XXVII*, Number 83, front page.

No byline (1921, October 27). Girl diver goes 162 feet down Simi well. *Oxnard Daily Courier*, p. 1. Retrieved from https://www.newspapers.com/image/31846714/

No byline (1921, December 15). Comment section. *Petroleum Age: The Oil Man's Business Paper, Vol. 8, No. 16*, p. 31. Retrieved from https://books.google.com/books

No byline (1922, February). Somebody said it couldn't be done. *The Rig and Reel magazine*. Source: Amelia Behrens Furniss's personal scrapbook. Copy provided courtesy of Noel Furniss.

Furniss, Noel, personal communication via e-mail January 5, 2016.

Transocean Personnel On board (POB) list (2010, April 10). Retrieved from http://www.mdl2179trialdocs.com/releases/release201303071500008/TREX-00687.pdf

Shand, J. (1968, December 11). Woman in a Man's World: The Oil-Hunt Girl. *Melbourne Sun*, p. 43. Copy provided courtesy of Dr. Marjorie Apthorpe.

Apthorpe, Dr. Marjorie, personal communication via e-mail August 6, 2018.

No byline given (2011, March 6). Oil boss and female role model. Aftenposten newspaper, Economy section. Translated from Norwegian to English by Google Translate. Retrieved from https://www.aftenposten.no/okonomi/i/JO18m/Oljeboss-og-kvinnelig-rollemodell

Cheang, C. C. (2015, October 26). PETRONAS' first female drilling supervisor settles in trailblazing role. *Rigzone*. Retrieved from https://www.rigzone.com/news/oil_gas/a/140964/petronas_first_female_drilling_supervisor_settles_in_trailblazing_role/?all=hg2

Aranha, J. (2017, August 3). Meet India's first woman firefighter, a trailblazer for generations of women to come! *The Better India*. Retrieved from https://www.thebetterindia.com/110666/first-woman-firefighter-india-harshini-kanhekar/

Nayak, P. (2016, August 3). What it takes to be the first woman firefighter of India – Harshini Kanhekar's story. Your Story. Retrieved from https://yourstory.com/2016/08/harshini-kanhekar/

Cuevas, G. (2012, May 21). My experience as a field engineer off the coast of Angola. Huffpo. Retrieved from http://www.huffingtonpost.com/women-20/women-engineers_b_1523304.html

Hegab, D. (2017, March). Breaking down stereotypes. *BP magazine*. Retrieved from https://www.bp.com/en/global/corporate/news-and-insights/bp-magazine/dena-hegab-egypt-female-engineer.html

Dreue, Anne-Christine, personal communication via e-mail June 9, 2016.

Clare, Melissa, telephone interview June 21, 2018, and via e-mail February 10, 2019.

Blum, J. (2018, August 2). Q&A: How OXY's CEO weathered the oil bust and came out stronger. *Houston Chronicle*. Retrieved from https://www.houstonchronicle.com/business/energy/article/Q-A-How-OXY-s-CEO-weathered-the-oil-bust-and-13126112.php

Akinbajo, I. (2015, October 10). Howard University faults Diezani Alison-Madueke's year of graduation. *Premium Times*. Retrieved from https://www.premiumtimesng.com/news/top-news/191295-throwback5-howard-university-faults-diezani-alison-maduekes-year-of-graduation.html

Viegas, N. (2016, May 06). Graca Foster passes almost unnoticed in law classes in Rio de Janeiro. *Época magazine*. Translated from Portuguese to English by Google Translate. Retrieved from https://epoca.globo.com/tempo/expresso/noticia/2016/05/graca-foster-passa-quase-despercebida-em-aulas-de-direito-no-rio-de-janeiro.html

Chapter 2 – Margaret McMillan

Guidry, Leigh (2016, September 2). Local 'legend,' offshore safety pioneer remembered. *Daily Advertiser*. Retrieved from https://www.theadvertiser.com/story/news/local/2016/09/02/local-legend-offshore-safety-pioneer-remembered/89766694/

McAllister, Glenda, personal communication via telephone July 25, 2015.

Maillet, Bonnie, personal communication via telephone July 17, 2015, and via e-mail November 9, 2015.

No byline (1977, December 4). Margaret McMillan retires from USL phys. education dept. *The Crowley Post-Signal*, p. 6. Retrieved from https://www.newspapers.com/image/470643572/?terms=Margaret%2B McMillan%2BRetires%2Bfrom%2BUSL%2BPhys.%2BEducation%2B Dept.

McMillan, Wikoff, personal communication via e-mail July 14, 2015.

No byline (1951, September 13). Eight SLI faculty members given leave of absence. *Daily World*, p. 16. Retrieved from https://www.newspapers.com/image/227252303/?terms=Daily%2BWor ld%2BEight%2BSLI%2BFaculty%2BMembers%2BGiven%2BLeave% 2Bof%2BAbsence

Domingue, Dave, personal communication via e-mail November 21, 2018.

McMillan, Margaret (2004, October 2). Offshore Energy Center Hall of Fame induction, pp. 6, 7, 11, 16. Interviewed by Dr. J. Pratt [in person, Houston, Texas]. (Transcribed copy provided by OEC.)

Landry, M. (2004, November 1). Lafayette woman praised for offshore safety work. *The Advocate*. Retrieved from https://athleticnetwork.net/site.php?pageID=54&profID=4433

McMillan, John, personal communication via telephone February 3, 2019.

McMillan, Wikoff, personal communication via e-mail February 7, 2019.

McMillan, David Haas, personal communication via telephone February 7, 2019.

Chapter 3 – Yassmin Abdel-Magied

Abdel-Magied, Yassmin, personal communication via telephone December 7, 2015.

Chapter 4 – Sara Akbar

Akbar, Sara, personal communication via telephone January 5, 2015.

Dundas, S., Picard, A., & Roberts, D. (Producers) & Douglas, D. (Director). (1992). *Fires of Kuwait* [documentary]. USA: IMAX Corp.

Curtiss, R.H. (1995, March). Giant-screen film records horrors of Kuwait's flaming oil. *Washington Report on Middle Eastern Affairs*, pp. 30, 106.

McKinnon, M. (Producer & Director). (1992). *Tides of War: Eco-Disaster in the Gulf* [documentary]. London: McKinnon Films Production for the National Geographic Society.

Chapter 5 – Jerry Tardivo Alcoser

Alcoser, Jerry Tardivo, personal communication via telephone July 23, 2015.

Chapter 6 – Ann Cairns

Cairns, Ann, personal communication via telephone September 14, 2015, and August 24, 2016.

Harlow, P. & McKenzie, S. (2015, February 18). Mastercard boss Ann Cairns' journey from oil rigs to credit card queen. *CNN Business*. Retrieved from http://www.cnn.com/2015/02/18/business/mastercard-boss-ann-cairns-oil-rigs-credit-card/

Chapter 6 – Sarah Helen Darnley

Darnley, Angela and Anne, personal communication via e-mail from July 9, 2015 to June 21, 2016.

No byline (2013, August 25). Helicopter crash victim's dad tells of frantic calls to his daughter after he hear about crash. Daily Record. Retrieved from https://www.dailyrecord.co.uk/news/scottish-news/helicopter-crash-victims-dad-tells-2218728

Nicolson, K. (2013, September 10). Family farewell to 'free spirit.' *The Press and Journal* (Moray), pp. 2 & 3. Retrieved from https://www.pressreader.com/uk/the-press-and-journal-moray/.../282900908268130

Chapter 7 – Myrtle Dawes

Dawes, Myrtle, personal communication via telephone August 24, 2015.

Adom, A. (2014, September). Me & my career. *Pride magazine*, pp. 120-121. Retrieved from www.pridemagazine.com MyrtleDawes_meandmycareer.pdf

Chapter 8 – Anne Grete Ellingsen

Ellingsen, Anne Grete, personal communication via telephone August 15, 2015.

Norwegian Petroleum Directorate (NPD) and Statoil (co-produced). (1985). *Norwegian Women Offshore* [documentary]. Norway. Retrieved from http://www.nb.no/kulturminne-statfjord/nb/ce575c7ba560e3dd6d7e060d04037175?index=11

Kammerzell, J. (2012, February 14). Anne-Grete Ellingsen: Elf's First Female Engineer. *Rigzone*. Retrieved from http://www.rigzone.com/news/oil_gas/a/115131/AnneGrete_Ellingsen_Elfs_First_Female_Engineer

Chapter 9 – Arlete Fastudo

Fastudo, Arlete, personal communication via e-mail March 21, 2016, and via
 Skype October 26, 2017.
Sonangol (2014, June). Making career waves. *Universo* magazine, pp. 35-37.

Chapter 10 – Abigail Ross Hopper

Hopper, Abigail Ross, personal communication via telephone August 19, 2015,
 and February 2, 2018.
No byline (Spring 2014). BOEM: protecting the oceans. *Winds of Change*.
 Retrieved from https://www.pohlyco.com/wp-
 content/uploads/2014/03/top50_hi.pdf

Chapter 11 – Eve Howell

Howell, Eve, personal communication via telephone November 10, 2015.

Chapter 12 – Dr. Amy Jadesimi

Jadesimi, Amy, personal communication via telephone February 10, 2016.
Al Jazeera. (2013, February 1). Tutu's children: the baton has been passed.
 (Episode 4.) [Video file.] Retrieved from
 https://www.youtube.com/watch?v=aLEpu_SA3jA

Chapter 13 – Zara Ibrahim Khatib, PhD

Legg, Corbett, personal communication via e-mail February 22, 2016, through
 May 25, 2016.
Legg, Caroline, personal communication via e-mail March 16, 2016.
Numerous friends' contributions personal communication via e-mail in 2016.

Chapter 14 – Las Mujeres en las Plataformas de Pemex

Ávila, Brenda Medina, personal communication via e-mail August 21, 2018.
Espinoza, Mary Betanzos, personal communication via e-mail August 21, 2018.
Ramón, María Franco, personal communication via e-mail August 21, 2018.
Rodríguez, Olga Cantú, personal communication via e-mail November 20,
 2018.
No byline. (2016, April 15). Rig workers: the soul of Mexico's crisis-hit oil
 sector. Agencia EFE. Retrieved from
  https://www.efe.com/efe/english/business/rig-workers-the-soul-of-
 mexico-s-crisis-hit-oil-sector/50000265-2898258
Gries, R. (2017). *Anomalies: Pioneering Women in Petroleum Geology 1917 –
 2017*. Denver, Colorado: JeWel Publishing, LLC.
Pemex and UNDP (co-produced). (2018). *For A Mexico With More Scientists,
 Engineers, and Mathematicians*. [Video files.] Retrieved from
 http://www.mx.undp.org/content/mexico/es/home/presscenter/articles/2

018/09/caja-de-herramientas---por-un-mexico-con-mas-cientificas--ingeni.html

Chapter 15 – Mieko Mahi

Mahi, Mieko, in-person interviews February 5, 2016, and July 21, 2018 and via e-mail December 30, 2018.

Ryan, C. (1959). *The Longest Day: The Classic Epic of D-Day, June 6, 1944.* New York: Simon & Schuster.

Economides, M. (2013, May 6). A profile of Mieko Mahi – photographer to the oil industry. *Fuelfix.* Retrieved from https://fuelfix.com/blog/2013/05/06/a-profile-of-mieko-mahi-photographer-to-the-oil-industry/

Chapter 16 – Alyssa Michalke

Michalke, Alyssa, personal communication via telephone September 11, 2015.

Chapter 17 – Deirdre Michie

Michie, Deirdre, personal communication via telephone November 6, 2015.

Chapter 18 – Scarlett Mummery

Mummery, Scarlett, personal communication via telephone June 29, 2015.

Viewpoints w/ Terry Bradshaw. (2014, November 21). *Tools and Technology Solutions Impacting Industry Success: Benthic & PROD* [video file]. Retrieved from https://www.youtube.com/watch?v=5jjwLyByuXs&feature=youtu.be

Chapter 19 – Jennifer DiGeso Norwood

Norwood, Jennifer, personal communication via telephone August 12, 2105.

Chapter 20 – Sue Jane Taylor

Taylor, Sue Jane, personal communication via telephone August 23, 2015, and via Skype April 15, 2018.

PwC (2015). *The female millennial: a new era of talent*, pp. 1-31, p. 24. Retrieved from https://www.pwc.com/jg/en/publications/the-female-millennial_a-new-era-of-talent.pdf

Hunter, Dr. James. (2013). Introductory essay to Beatrice Works exhibition. Aberdeen Art Gallery and Museums Publication.

Taylor, S. J. (2005). *Oilwork North Sea Diaries 1984 – 2004* (Birlinn Publication; 2005), Chapter 7 Piper Alpha Platform North Sea, p. 106. Chapter 12 Piper Bravo Platform 2004, p. 171. Reprinted with permission from the author. All rights reserved. ©Sue Jane Taylor.

McGinty, S. (2008, December 1). *Fire in the Night: The Piper Alpha Disaster.* (London: Palgrave McMillan, Ltd.).

Taylor, S. J. (2009). *Mineworks, Underground Tour, Mount Lyell copper mine.* Reprinted with permission from the author. All rights reserved. ©Sue Jane Taylor.

Taylor, S. J. (2005). *Oilwork North Sea Diaries 1984 – 2004* (Birlinn Publication; 2005), Author Sue Jane Taylor, Chapter 11 Piper Alpha Platform North Sea, page 126. Reprinted with permission from the author. All rights reserved. ©Sue Jane Taylor.

Chapter 21 – Pat Thomson

Thomson, Pat, personal communication via telephone May 16, 2015, and in-person interview October 8, 2015.

Thom, Dr. Alix. Oil & Gas UK (2014), UK Continental Shelf (UKCS) *Offshore Workforce Demographics Report.* Retrieved from https://oilandgasuk.co.uk/offshore_workforce_demographics/?_cldee=c GF1bC5nYXlAaW1sZ3JvdXAuY28udWs%3D

Laroche, J-P (executive producer). Rombout, R. (director). (1994). *Black Island* [documentary]. Wallonie Image Production - BRTN Dienst Cultuur - Galatée Films - Pandora Productions.

Chapter 22 – Nina Vorderwülbecke

Volderwülbecke, Nina, personal communication via telephone October 8, 2015.

Chapter 23 – Marni Zabarsky

Zabarsky, Marni, personal communication via telephone July 29, 2015, and November 4, 2017.

No byline. (1979, Winter). *Keep on Trukken.* Faceplate: the official magazine for the divers of the United States Navy. Volume 10, No. 4, pp. 18-19. Retrieved from http://navydiverfoundation.org/assets/1979_winter.pdf.

Legacy Capital. (2006, March 9). Press release. Retrieved from http://www.legacycapital.com/portfolio/epic-divers-marine-epic-has-sold-its-assets-and-business-to-tetra-technologies-inc/

Chapter 24 – Alyssa Michalke

Michalke, Alyssa, personal communication via telephone September 11, 2015.

Afterword

Spitzmueller, C. and Brown, H. (2019, February 1). Not enough talent for the energy workforce? Energy's diversity problem may be the solution. Forbes. Retrieved from https://www.forbes.com/sites/uhenergy/2019/02/01/not-enough-talent-for-the-energy-workforce-energys-diversity-problem-may-be-the-solution/#744a0a9a3f97

About the Author

Rebecca Ponton is an American journalist. Her work has appeared in numerous American and international publications, such as *National Geographic Traveler, CNN Traveller, Texas Monthly, AAA Texas Journey, MORE, Ms. Magazine, Arabian Woman, Emirates Woman, Dubai Airport Magazine, Gulf Air,* and others. She is also a petroleum landman and obtained her certification through the University of Texas Petroleum Extension (PETEX). As an expatriate from  1992 – 2006, she lived in Dubai and Abu Dhabi, United Arab Emirates, Baku, Azerbaijan, and Almaty, Kazakhstan. During her time in Azerbaijan, she wrote two booklets for the United Nations. This is her first book and the first in the Breaking the GAS Ceiling™ series. She received her Bachelor of Arts English from the University of Texas – San Antonio (UTSA). A native Texan, originally from Corpus Christi, she currently lives in San Antonio.

Rebecca can be reached at BreakingTheGasCeiling@yahoo.com. Further details about the book can be found at www.BreakingTheGasCeiling.com.

♀ ♀ ♀

Index

CPSIA information can be obtained
at www.ICGtesting.com
Printed in the USA
LVHW052125240419
615446LV00001B/1/P